J. D. (Josiah Dwight) Whitney

Names and Places

Studies in Geographical and Topographical Nomenclature

J. D. (Josiah Dwight) Whitney

Names and Places
Studies in Geographical and Topographical Nomenclature

ISBN/EAN: 9783743418721

Manufactured in Europe, USA, Canada, Australia, Japa

Cover: Foto ©berggeist007 / pixelio.de

Manufactured and distributed by brebook publishing software (www.brebook.com)

J. D. (Josiah Dwight) Whitney

Names and Places

NAMES AND PLACES

STUDIES IN GEOGRAPHICAL
AND
TOPOGRAPHICAL NOMENCLATURE

By J. D. WHITNEY

CAMBRIDGE
PRINTED AT THE UNIVERSITY PRESS
1888

One Hundred Copies Printed.

CONTENTS.

	PAGES
APPALACHIAN AND CORDILLERAN	7–27
OREGON AND PEND' OREILLES	28–75

TOPOGRAPHICAL NOMENCLATURE.

I. INTRODUCTORY	76–82
II. MOUNTAINS, PEAKS, AND SIERRAS	83–128
III. VALLEYS, GORGES, AND CAÑONS	129–172
IV. PLAINS, PRAIRIES, AND SAVANNAS	173–234

INDEX OF TOPOGRAPHICAL NAMES . 235–239

STUDIES IN GEOGRAPHICAL

AND

TOPOGRAPHICAL NOMENCLATURE.

APPALACHIAN AND CORDILLERAN.

I. APPALACHIAN.

THE Appalachian Mountains were first seen by Cartier, in 1535, who, when navigating the St. Lawrence, had in sight the portion of the range which extends to the south of that river, through what is now the State of Maine. De Soto in his explorations (1538-1543) became acquainted with this mountain system, around the southern extremity of which he made his way to the southwest and thence up the Mississippi River. This bold explorer first gave currency to the name "Apalache," which also appears under the form of Palassi (Montagnes de Palassi) in the report of Laudonnière's expedition of 1564. This name was one given by the aboriginal inhabitants, who also furnished Laudonnière with a specimen of native gold from a region at the base of the Appalachians where large quantities of the precious metal have since been obtained. The name "Apalache" appears on Mercator's map of 1569 in the form of "Apalchen;" and the delineation of this chain of mountains on this map is approximately correct, in so far as it indicates a range extending parallel with the coast, through what is now the United States, and bending to the east in the northern portion (Norombega) parallel with the St. Lawrence.

On one of the maps accompanying De Laet's "Novus Orbis seu Descriptionis Indiæ Occidentalis Libri xviii." (1633), a group of hills is indicated surrounding a small lake, cut by the parallel of 35°; and to these hills the name "Apalatcy Montes" is given. The name "Apalache" is also found on this map a little farther west of the Apalatcy Montes, apparently intended to indicate the position of a region inhabited by a tribe of that name, since in the text, in a summary of the discoveries of Panfilo de Narvaez, in speaking of nuggets of gold given to this explorer by the natives, they are described as "auri ramenta aliquot, quæ barbari ab *Apalache*, longissimo intervallo ab ipsis dissita et auri divite regione, se habere testabantur." The word "Apalache" seems therefore, beyond doubt, to have come from the southern part of the region now called Appalachian, and to have been the aboriginal name of a locality, or of a tribe of Indians inhabiting that portion of the country.

The first recognition of the peculiar topographical character of the Appalachian system, and, indeed, the first important approximately correct map of any portion of the interior of the United States, is due to Lewis Evans, the first edition of whose map bears the date of 1749. On this map, which is entitled "A Map of Pennsylvania, New Jersey, and New York, and the Three Delaware Counties," the Appalachian Mountains are indicated as being made up of a number of distinctly parallel ranges, and in accompanying "remarks" engraved upon the map the principal topographical features of the system are described with remarkable perspicuity and insight into their character. These mountains are said by Mr. Evans to be "not confusedly scattered here and there in lofty peaks overtopping one another, but stretching in long uniform ridges, scarce half a mile perpendicular in any place I saw them." In another map by the same geographer, entitled "A General Map of the British Colonies in America," bearing date of 1755, and to which thirty-two quarto pages of text are appended, the Appalachian Mountains are indicated with more detail, and as being made up of a much larger number of subordinate ranges than are shown on the map of 1749. On the map of 1755 the main subdivisions of the Appalachian System, as now recognized, are clearly outlined by Mr. Evans, and a distinct name given to each. The South Mountains and the continuation of them under the name of the Blue Ridge are laid down; the Alleghany Ridge (spelled by Mr. Evans Alle-géni) is shown; and the Appalachian subdivision of the Appalachian System, as this term was used by the First Pennsylvania Geological Survey, also receives a distinct name — the Endless Mountains — and is described in considerable detail with a remarkable comprehension of its principal topographic features. The name "Endless" is said to be "a translation of the Indian name bearing that signification." This name "Endless," adopted or suggested by Mr. Evans, seems not to have met with approval, never having become current; but most of the other names on his maps are those still in use, while some have entirely disappeared.

On the map of the southern portion of North America compiled by Poirson from the materials collected by Humboldt on his American journey, and published in 1811, a continuous lofty range of mountains is indicated as extending from Alabama to New York, the southern portion of which is designated as the "Montagnes Apalaches ou Alleghany" while the northern extension is called "Mont Alleghany," thus indicating as existing at that time a condition of things which continued for many years later; namely, an uncertainty as to whether the complex of ranges in question should bear the name of Apalachian (Appalachian) or Alleghany. Morse, the earliest American geographer, in his Gazetteer (third edition, 1810) says as follows: "The general name of the whole range, taken collectively, seems to be undetermined. Mr. Evans calls them the Endless Mountains; others have called them the Appalachian Mountains, from a tribe of Indians who live on a river which proceeds from this mountain, called the Appalachicola; but the most common name is the Alleghany Mountains, so called probably from the principal ridge of the range."

On Maclure's Geological Map of the United States, accompanying an article read before the American Philosophical Society in 1817, and published in 1818, the name "Alleghany," spelled "Allegany," is repeated three times, but always on the westernmost member of the system of ranges, while various other designations are applied to more eastern portions of the system. Some of these designations appear never to have become current; others, like that of the Blue Mountains, are still, and have — since the time of Evans, at least — been in use. Maclure's map emphasizes a tendency existing from early days to limit the use of the name "Alleghany" to that portion of the Appalachian Range which in the form of a bold escarpment marks, in Pennsylvania, the most western important topographical feature of the region, and from the crest of which to the west declines a gently rolling plateau-like region, uninterrupted by conspicuous ridges.

At the time of the beginning of the Geological Survey of Pennsylvania (1836), when the necessities of a more precise geographical nomenclature began to be felt, the State geologist, H. D. Rogers, wished to retain the name "Alleghany" (spelled by him "Allegheny") for the escarpment and plateau for which this name was most current, and also to limit the term "Appalachian" to the middle area of Pennsylvania, comprising the "wide mountainous zone embraced between the southeastern region [the South Mountains and Blue Ridge] and the principal ridge of the Allegheny Mountains." In reference to this matter of nomenclature, Professor Rogers says, in his First Geological Report (1836) : " It is true that the signification of the word ['Appalachian'] has been so extended as to comprehend the great table-land of the Allegheny Mountains and its spurs, though it is greatly to be wished, for the sake of giving greater exactness to geographical reference, that the latter mountain, with the ridges west of it, should be known exclusively as the Allegheny chain; and the mountains from its base, east, to the great Cumberland Valley, exclusively as the Appalachian chain. When occasional allusion may be necessary to these two systems of mountains, so dissimilar both in their geological structure and their external configuration, we shall always employ the names 'Allegheny' and 'Appalachian' in the restricted sense here specified."

In spite of the manifest desire of the geologists of the First Pennsylvania Survey to limit the name " Appalachian " to a portion of the range, the need of some general designation for the whole region becoming more and more decidedly felt, and such not having been furnished by Professor Rogers, his wishes and those of his assistants have been disregarded; and during the past twenty years, by general usage, the entire system of ranges from Gaspé to Georgia, with all its valleys and table-lands, has become known to physical geographers and geologists as the Appalachian Range or System. No doubt the fact that Guyot's all-important paper, published in 1861, bore the title "On the Appalachian Mountain System," had a marked influence in favor of bringing about the present unanimity of usage in this matter among geographical and geological writers.

That Guyot was himself somewhat in doubt whether to call this system of ranges "Appalachian" or "Alleghany," is evident from the fact that on the map accompanying this paper, which was engraved and published in Germany a year before its appearance in America, the latter name is used. This map, in both the German and American editions, has as its title "Physikalische Karte des Alleghany-systems."

II. CORDILLERAN.

THE mountainous region on the western side of the continent remained almost a *terra incognita* until after the beginning of the present century. In spite of the known great breadth of the continental mass in the latitude of the United States, and notwithstanding the fact that as early as the middle of the sixteenth century ranges of mountains had been seen on the Pacific coast by explorers, notably by Cortez and Cabrillo, and named by them, while subsequent explorations down to the time of those of Vancouver made the public aware that there were high mountains on the western side of the continent, the fact that these were a portion of an immense complex of ranges, valleys, and table-lands seems hardly to have become appreciated by geographers until nearly the middle of the present century.

As late as the year 1794 the text-book of geography chiefly, if not exclusively, used in the United States (Morse's "Geography made Easy," fourth edition), contained the following statement repeated from former editions: "North America, though an uneven country, has no remarkably high mountains. The most considerable are those known under the general name of the *Allegany Mountains*. These stretch along in many broken ridges under different names from Hudson River to Georgia. The *Andes* and *Allegany Mountains* are probably the same range interrupted by the Gulf of Mexico."

In 1802 there seems to have been the first recognition, on the part of American geographers, of the fact of the existence of mountains on the western side of the continent which were in some sort the continuation of the already somewhat familiar Andes of South America. In Morse's "American Universal Geography," fourth edition, 1802, we read as follows: "In New Spain the most considerable part of this chain [the Andes] is known by the name of Sierra Madre. . . . Farther north they [the ranges of the Sierra Madre] have been called from their bright appearance the *Shining Mountains*." Again, in the fifth edition of the "Elements of Geography" by the same author, published in 1804, we have the same statement repeated in regard to the "Shining Mountains," with the additional notice that they lie "away west of Louisiana," and are but little known.

At the beginning of the present century the names "Shining" and "Stoney," or "Stony," were both given by different geographers and cartographers to a range of mountains on the western side of the continent, which was indicated on various maps as a single ridge, and placed in various positions, sometimes in the vicinity of the one hundredth meridian and sometimes on the remotest northwestern edge of the continent. Thus Arrowsmith's map of North America (1795) has the name "Stony Mountains" upon it, with the remark that "they are 3,520 feet above the level of their base, and according to the Indian accounts of five ridges in some parts." But a later edition of this work — that of 1802 — has "Rocky" in the place of "Stony" Mountains. The name "Mountains of the Shining Stones" is also used on various maps issued towards the close of the eighteenth century; also "Mountains of Bright Stones," which latter is the name found on the map accompanying Carver's Travels, and which bears the date of 1778. In the text, however, these mountains are called the "Shining," and the origin of this name is thus stated by him: "Among these mountains, those that lie to the west of the River St. Pierre are called the 'Shining Mountains,' from an infinite number of chrystal stones of an amazing size, with which they are covered, and which when the sun shines full upon them sparkle so as to be seen at a very great distance." Carver seems to have been the first to use the names "Shining Mountains" and "Mountains of the Bright Stones." The locality where the sparkling crystals occur which so excited Carver's imagination, is not known; but it has been suggested by the Abbé Domenech that the crystalline plates of selenite in the Tertiary beds of the Bad Lands may have been the "Shining Stones" which gave the name for a time to the mountainous region to which the Bad Lands form a sort of introduction, and which had at a very early period become known to the fur-hunters and trappers of the Far West.

Although the name "Rocky Mountains" is the one exclusively employed by Lewis and Clarke in the report of their expedition (1804–1806), yet the term "Stony," which was that used by Jefferson in his instructions to them, continued to make its appearance on the various maps issued by Morse up to as late, at least, as 1812 (American Universal Geography, sixth edition). Gradually, however, "Rocky" took the place of "Stony;" and "Shining," as a name for the complex of ranges, or any part thereof, soon entirely disappeared from the map. Long after the time of Lewis and Clarke, however, various attempts were made to give entirely new names to these mountains; as, for instance, by Tardieu in his finely engraved map of Louisiana and Mexico, published at Paris in 1820, in which the main range, forming the back-bone of the Far West, is, in three places, named "the Columbians [*sic*] Mountains," while the designation "Rocky" is limited to a small spur or parallel range on the east, occupying a very subordinate position as compared with that assigned to the "Columbians." Again, in the geological map and sections accompanying the English edition of Hinton's "History and Topography of the United States," of which a new and improved edition was published in Boston in 1834, the Rocky Mountains are called the "Chippewayan Mountains."

All the older maps are defective, especially in that they do not recognize the fact that the western highlands are made up, not of one, but of many ranges, quite distinct from each other, and often separated by wide valleys and table-lands. The map accompanying Humboldt's "New Spain," previously alluded to as having been compiled by Poirson, was the first attempt to make it appear that the orography of the western region was much less simple than it had been previously assumed to be. On this map a very marked mountain range is indicated as closely bordering the Pacific, and continuous from the southern extremity of Lower California to the northern limit of the map in latitude 42°. Another very strongly indicated continuous range extends through the centre of Mexico, north through what is now the United States as far as the northern limit of the map. This range is placed approximately on the meridian of 109° to 110° (west of Paris). Behind the Pacific Coast Range thus represented, there is in the latitude of Central California a vague indication of another range lying farther eastward; to this the name of "Sierra San Marcos" is given, and this may be taken as a hint at a recognition of the existence of the Sierra Nevada, although the range is placed much too far from the Pacific. The remainder of the area between the two enclosing ranges, to neither of which is any general designation assigned, is occupied with a few vague and incorrect details, but more nearly correct than that which is given on most of the maps published during the twenty years following the appearance of Humboldt's map, since in these the streams running into the Pacific are made to head far to the east, in what would now be designated as the Rocky Mountain Range proper, and to run almost due west to the sea. On Humboldt's map different portions of the range forming the eastern boundary of the Cordilleran region are designated as the Sierra de los Mimbres and the Sierra de las Grullas. A part of the Pacific Coast Range is called the Sierra Santa Lucia, a name still current; and the extreme northern portion of the Coast Range bears the name of Sierra Nevada, which is that now given to the range next east of the San Joaquin and Sacramento Valleys.*

* This name first appears on the map made after Verrazano's chart, by Michael Lok, published in 1582, and republished by the Hakluyt Society in 1850. It is spelled "Sierre Neuada," and designates a range of mountains running east from the head of the "Mare Bermejo," and forming the northern boundary of the continent along an extent of 15° of longitude.

The expeditions of Bonneville (1832-1836) and of Fremont (1842-1844), which made known the existence of an interior closed basin in the western highlands, and also revealed the principal features of the topography of that region, soon followed as they were by the discovery of gold in the Sierra Nevada, almost immediately made this designation a familiar one all over the world; and with the adoption of this name for the prominent and important range on the western edge of the country, it was natural that the designation of "Rocky Mountains" should become more and more limited to the eastern edge of what was gradually becoming recognized as being a great complex of ranges, of which, however, the limiting ones on the east and west were on the whole the most elevated and continuous. Meanwhile the interior ranges, or those enclosed between the Rocky Mountains and the Sierra Nevada, received names, or retained those already given by the aboriginal inhabitants, so that by the time the Pacific Railroad was completed nearly all the ranges had distinctive appellations. Still, up to quite a recent date there was no collective name for the whole system of ranges on the western side of the continent, including the Rocky Mountains proper on the east and the Pacific Coast Ranges on the west, and all the region of mountains, table-lands, and valleys enclosed between them. The desirability of such a name became, however, more and more manifest, as the region in question began to be written about as a whole, and to be recognized as forming one all-important feature in the topography of the country. The condition of things in regard to the nomenclature of the western complex of ranges was almost exactly what it had been in the east before the name "Appalachian System" or "Appalachian Range" had been adopted by geographers as a designation for the highlands of the eastern side of the continent.

In 1868 the present writer, in a work devoted to a topographical description of a portion of the Sierra Nevada (the Yosemite Book), suggested the use of the name "Cordilleras" (with the adjective "Cordilleran"), as a proper, convenient, and euphonious designation of the great western complex of ranges; and this name has been generally accepted and made use of, especially in the various publications of the Census Bureau, including those of the Census of 1870 and 1880. A few words, however, may here be added in reference to the origin of this designation and the convenience and propriety of its use in the manner designated.

Before any definite knowledge of the mountainous region on the western side of North America had been obtained, considerable progress had been made toward a clear understanding of the nature and extent of the clearly defined and lofty ranges on the Pacific side of the southern division of the American continent. The journeys of Humboldt in that region, and the voluminous publications following the completion of these explorations, were the principal cause of this condition of things; but the great simplicity of the orographic structure of South America as compared with the complexity of the northern topography, and especially of that portion within the limits of the United States, was an additional reason why the geography of this latter region was so slowly worked out.

Humboldt, in his "Personal Narrative" of his South American travels, uses the term "Cordilleras of the Andes" as the most general designation for the system of ranges extending from Patagonia along the Pacific coast "to the mountains lying at the mouth of the

Mackenzie River." In this work he sometimes calls the South American division of the Cordilleras simply the Andes, and sometimes the Cordillera (and also Cordilleras) of the Andes; the prolongation of these ranges to the north of the Isthmus is designated by a variety of names, sometimes as the Andes of New Mexico, sometimes as the Cordilleras of Mexico, and sometimes the Andes of Anahuac. The most western division of the Cordilleras he usually calls the Maritime Alps, and occasionally the Mountains of California. That the topography of the western side of the North American continent was but vaguely and imperfectly known at this time, is indicated by the fact that Humboldt prolongs the eastern chain of the Andes from Potosi through Texas and the Ozark Mountains to the Wisconsin Hills, of which he says: "Their metallic wealth seems to denote that they are a prolongation of the eastern Cordillera of Mexico." In Humboldt's last great work, "Kosmos," written soon after the results of the Pacific Railroad surveys had become known, he uses the name "Cordilleras" as the equivalent of the Andes, understanding by it, in general, the Andes of South America; sometimes, however, including the mountains of Mexico. For the continuation of these ranges farther north the name generally adopted by him is "Rocky Mountains" (Felsengebirge), and instead of "Maritime Alps" he uses the names, already current in the United States, of "Sierra Nevada" and "Cascade Range."

From Humboldt's time on, however, the name "Andes" became more and more limited in its use to South American ranges, and the word "Cordilleras," * which simply means "mountain ranges," was omitted, so that a long time has now elapsed since geographers began — as a general rule — to designate the South American Pacific coast mountain system as simply the Andes, that term having been entirely dropped as a name for any part of the North American ranges. This condition of things seems to leave the term "Cordilleras" as a convenient, suitable, and euphonious one for designating the entire complex of western North American ranges, although it was not, as some have stated, thus used by Humboldt, who never at any time proposed any general designation for the northern division of his "Cordilleras of the Andes," as is clearly evident from what has been stated above. That the introduction of the name "Cordilleras" with

* See further on in this volume (p. 88).

the meaning given to it by the present writer met a distinctly felt want among geographers, is sufficiently proved by its immediate adoption by authors of important works in which the general topography and geology of North America came under discussion.

OREGON AND PEND' OREILLES.

> "Where rolls the Oregon, and hears no sound
> Save his own dashings."

THERE are two geographical names in North America which have given rise to much speculation as to their origin and derivation, with as yet but very unsatisfactory results. These names are Canada and Oregon. The present communication will, as the writer believes, settle the difficulty with regard to the second of them, although it is possible that the Oregonians themselves may not be particularly well pleased with the solution.

The first appearance in print of the name "Oregon" was in the "Travels through the Interior Parts of North America in the Years 1766, 1767, and 1768," by J. Carver, Esq., from which (second edition, published at London, 1779) the following extract is quoted: "The four great rivers that take their rise within a few leagues of each other, nearly about the centre of this great continent; viz.

the River Bourbon, which empties itself into Hudson's Bay; the waters of Saint Lawrence; the Mississippi, and the River Oregon, or the River of the West, that falls into the Pacific Ocean at the Straits of Annian."

Carver furnishes no explanation of the origin or meaning of the name "Oregon." Besides its occurrence in the quotation above given, the name (spelled "Oregan") is once more used by him (p. 542) when this river is again called the "Oregan or River of the West." On the general map attached to this work the name "Oregon" does not appear, but only that of "River of the West," the head of which is laid down in about longitude 104° and latitude 46°, and the mouth in latitude 45°, with the legend "discovered by Aguilar." The course of this river as given by Carver on this map is an exact copy of the same on a map of J. Palairet, revised by L. Delarochette, and published in 1763. Carver did not himself see any part of the river which he called the Oregon, or any of its tributaries; but it seems that soon after the appearance of his book this name began to come into use. Had it not been picked up by him, it might never have become known beyond the region where it originated. Carver's representation of the country to the north and west of Lake Superior, as far west as the "Western Sea," as well as that of the Pacific coast, is an almost exact copy of Palairet, the legends only being somewhat altered, while his use of that map in preparing his own is also clearly shown by the fact that the scales of the two are exactly the same. These remarks apply to that one of the two maps accompanying Carver's travels which bears the title of "A New Map of North America from the Latest Discoveries, 1778. Engraven for Carver's Travels." The other map in the same volume, entitled "A Plan of Captain Carver's Travels in the Interior Parts of North America in 1766 and 1767," embraces Lakes Superior and Michigan and the region west, extending a little way beyond the Mississippi with the Lake of the Woods in the northwestern corner. It has the author's route upon it, and is somewhat different in its details from the other map in the same volume, which embraces all of North America lying between the parallels of 25° and 70°. Directly south of the Lake of the Woods, on the parallel of 47°, on Carver's route-map, are the words "Heads of Origan."

It is very easy to make out whence Carver got the idea of a "River of the West." During the French occupation of Canada, from the early part of the last century on, there was constant talk among the explorers about the possibility of reaching the "Sea of the West" by some interior route starting from the Great Lakes. Since the French carried on their explorations almost exclusively by the aid of canoes and boats, they naturally followed the streams and lakes, in this respect forming a marked contrast to the Spaniards, who spread themselves on horseback towards the northwest, introducing the horse among various Indian tribes far down the Columbia, to such an extent that Lewis and Clarke, through much of the region they traversed, found that the aborigines were as familiar with this animal as the Spaniards themselves. Indeed, no inconsiderable part of the work of these adventurous explorers was done by the aid of horses purchased from the Indians.*

* "The horse is confined principally to the nations inhabiting the great plains of the Columbia, extending from latitude forty to fifty north, and occupying the tract of territory lying between the Rocky Mountains and a range of mountains which pass the Columbia about the Great Falls from longitude 16° [*sic*, in both English and American editions, — should be 116°] to 121° west. The Shoshonees, the Chopunnish, Sokulks, Escheloots, Eneshures, and Chilluckittequaws, all enjoy the benefit of that docile, noble, and generous animal; and all of them, except the last three,

The French, on the other hand, were as naturally led to extend their trading-posts, and to explore geographically the region lying to the north and northwest of Lake Superior, towards Lake Nepigon, the Lake of the Woods, Winnipeg, and the Assiniboine River. Explorations in this direction took them far away from the continental divide, which they must have crossed before they could find waters flowing into the Pacific.

Still, the idea of reaching the Sea of the West from the east through a connected chain of rivers and lakes was a favorite one with the Canadian explorers, both lay and missionary, among whom certainly were men of unbounded zeal and abundant courage. Based on various stories gathered by the traders from the Indian tribes to the west of the Lake of the Woods and the Assiniboine, the idea of a "River of the West" leading to the "Sea of the West" gradually took form, and began to be made a feature of the maps furnished

possess immense numbers." — LEWIS AND CLARKE, Eng. ed., p. 461.

"The Chayennes reside chiefly on the heads of the river [Chayenne], and steal horses from the Spanish settlement, a plundering excursion which they perform in a month's time." — LEWIS AND CLARKE, Am. ed., vol. i. p. 95.

by various explorers. In one of the maps accompanying the "Histoire et Description Générale de la Nouvelle France," by Charlevoix, we find the notion of a River of the West completely developed. This map bears on its face the legend "dressée par N[icolas] B[ellin] Ing. du Roy, et Hydrog. de la Marine" (1743). It shows a complete waterway, consisting of alternate river and lake, through from Lake Superior to about longitude $135°$ (west of Paris), where "suivant le raport des Sauvages commence le Flux et Reflux."

In a later map of Bellin's, published separately, and bearing the title of "Carte de l'Amérique Septentrionale depuis le 28 Degré de Latitude jusqu'au 72" (1755), the geography of the Northwest is very considerably changed, and in some respects improved. Lake of the Woods is moved ten degrees nearer to its true position. Lakes Winnipeg and Winnipegosis are better defined, but still greatly too far to the west. The "River of the West" disappears; while along the course of the Assiniboine (Riv. des Assiniboiles), which is correctly given as uniting with the Red River just before it falls into Lake Winnipeg, it is noted "qu'on peut croire aller à la mer de l'Ouest."

A few pages may here be devoted to an examination of what was done by the French during their occupation of Canada in the way of geographical exploration of the region lying to the west and northwest of the Great Lakes.*

D'Iberville, in 1700, proposed an exploration of the region beyond the Mississippi to the southwest. In a note developing this idea he says: "Trouvant la hauteur des terres et les rivières qui descendent à la Mer de l'Ouest, sçavoir si on les descendra, si elle tombe dans la Californie près des establissemens des Espagnols, et sçavoir s'il y aura seureté de s'aller livrer à eux ou aux Sauvages qui leur sont soumis."

Again, in 1703, the same officer, in a letter dated at La Rochelle, speaks of twenty Canadians having started from Tamaroas for New Mexico, "pour y commercer des piastres et voir ce que sont les mines dont les Sauvages leur ont parlé." In 1705 "un nommé

---

* The facts here stated in regard to the various attempts of the French to reach the Sea of the West, or to explore the region at the head of the Missouri, are compiled from the original documents published in the sixth part of Margry's "Découvertes et Établissements des Français dans l'Ouest et dans le Sud de l'Amérique Septentrionale" (1614–1754). Mémoires et Documents originaux: Paris, 1886.

Laurain" brings to the Chevalier de Beaurain news about the Spaniards living on the borders of New Mexico. In 1708 La Salle plans an expedition to discover the source of the Missouri. He says, in giving reason why this should be attempted, that he has learned positively "que des hommes blancs, comme nous, qui ne sont autres que les Espagnols, vont fort fréquemment avec des mulets en ce pays. . . . . Il y a des voyageurs canadiens qui l'ont remontée presque 3 à 400 lieues au nord-ouest et à l'ouest, sans qu'ils aient pu apprendre d'ou provient la source." In 1709 a French officer named Mandeville reports that the exploration of the Missouri would lead to great discoveries. In 1714 the Missionary Lemaire says, " On a remonté la Missouri plus de 400 lieues sans rencontrer aucune habitation Espagnole, et ce n'est qu'à quelque 500 lieues qu'on commence à en avoir des nouvelles par des Sauvages, qui font la guerre avec eux."

An elaborate memoir addressed to the "Conseil de la Marine" in 1717, by the Sieur Hubert, sets forth the advantages which would accrue to the Government of France from the exploration of the Upper Missouri. In this memoir the mines worked by the Spaniards, already several times mentioned in despatches of previous years, are again brought forward, and reason given why the high mountains at the head of this river might be expected to prove metalliferous. The point is raised whether, if the Spaniards should be found to be in possession of valuable mines in that region, it might not be possible to drive them out with the aid of the natives, "qui ont les Espagnols en horreur, et qui seroient dans les interests des François." Furthermore, it is suggested that at the source of the Missouri there will be found "une grande rivière qu'on prétend qui sort de la mesme montagne ou est la source du Missouri. On croit mesme qu' c'en est une branche qui va tomber dans la Mer de l'Ouest." This seems to be the first time the idea of a "River of the West" was prominently brought forward in any official despatch from the French in Canada to the home Government, and it is strongly insisted on that by this route commercial relations could be opened with Japan and China, — "le chemin en serait court." To this is added: "Cela paroist d'une grande importance à mériter d'en approfondir la vérité."

In a memoir signed "Bégon," added to a letter of Messrs. Vaudreuil and Bégon, bearing date Nov. 12, 1716, the statement is made that twenty-eight years before — in 1688, namely — the Assiniboines had offered to conduct a traveller named De Noyon to the Sea of the West, on the borders of which people went on horseback. The journey thither and back was to occupy five months; the river (Ouchichiq, "which leads to the Lac des Assiniboils [Manitoba and Winnepegosis] and thence to the Sea of the West") is said to be "très belle," and tide-water ("le flux et reflux") would be met at three days' journey from the sea. It is insisted at the end of this memoir that in order to derive any profit from commerce with the Sea of the West, it must be by means of the land route indicated.

In view of the great expense of the explorations thus meditated, it was decided to wait and endeavor to acquire more definite information in regard to the Sea of the West; and it was for that purpose that the services of Charlevoix were engaged, who in 1723, after three years of exploration and inquiry, reported that the best way to reach the Sea of the West was to ascend the Missouri; but this plan was definitely rejected by the Government, and it was concluded, instead of carrying out the ideas of Charlevoix, to establish missions among the Sioux, which was done on the borders of Lake Pepin, where, from the year 1723 on, a post was irregularly maintained and finally definitely abandoned by Legardeur de Saint Pierre in 1737.

A few years later the name of Pierre Gauthier de la Verendrye became conspicuous in the French Archives as that of an explorer strongly possessed with the idea of reaching the Sea of the West. He was a man of zeal, of some means, and the father of three sons who helped him in his task, and one of whom was killed in the course of his explorations. De la Verendrye, in a memoir addressed to the Minister of the Marine, dated Oct. 31, 1744, sets forth that for thirteen years he has been engaged in endeavoring to reach the Sea of the West, during which time he has suffered the greatest hardships. He had established trading-posts on Rainy Lake, Lake of the Woods, Lake Winnipeg, the Assiniboine River, and the Saskatchewan (Poskoyac). In 1742 he sent two of his sons with a well-equipped party " aux Mantanes,"— that is, to the region inhabited by the Mandans, — with the idea of penetrating in a southwesterly direction to the mountains, and reaching the much-wished-for Western Sea. They returned after fifteen months of absence, and the report of their journey was forwarded by the Governor-General of Canada (De Beauharnois) to the Minister, Oct. 27, 1744. Unfortunately this report is so vague that it seems impossible to make out much more from it than that the expedition came within sight of the Rocky Mountains in January, 1743, having left the Mandans July 23, and having travelled in general in a southwesterly direction, probably in the wide belt of land lying between the Yellowstone and the Missouri. The French were accompanied for some time by a large band of Indians ("Gens de l'Arc"), and with them were obliged to turn back, when near the base of the mountains, on account of their fear of the "Gens du Serpent." At the time of this turning back they may — as nearly as can be guessed — have been within one or two hundred miles of Snake River, and distant fifteen degrees of longitude from that Western Sea, of which La Verendrye hoped to get a sight from the summit of the mountain at whose base he was. Here they heard accounts of the missions of the Spaniards in California, which contained enough of truthful items to prove beyond doubt that there had been communication across the country between the Pacific coast and the Upper Missouri Region.*

The maps accompanying the work of Charlevoix have already been noticed. This work was published at just about the same time that the expedition of La Verendrye took place. Indeed, its preface bears a date very near that of the day of the return to Fort la Reine of La Verendrye's expedition, July 2, 1743. How confident Bellin, the cartographer who prepared the maps accompanying the narrative of Charlevoix, was, that the Sea of

---

* The following is a quotation from the report of La Verendrye to De Beauharnois in regard to this journey. It contains a part of what the old chief of the "Gens de l'Arc" told him in regard to the people living on the ocean, and whom they were prevented from reaching by fear of the "Gens du Serpent."

"Il poursuivait son discours ainsi : 'Les François [*Spaniards*; see further on] qui sont à la mer, me dit-il, sont nombreux ; ils ont quantité d'esclaves, qu'ils établissent sur leurs terres dans chaque nation ; ils ont des appartements séparés, ils les marient ensembles et ne les tiennent pas gênés, ce qui fait qu'ils se plaisent avec eux et ne cherchent pas à le sauver. Ils élèvent quantité de chevaux et autres animaux, qu'ils font travailler sur leur terre. Ils ont quantité de chefs pour les soldats, ils en ont aussi pour la prière.' Il me dit quelques mots de leur langue. Je reconnus qu'il me parloit Espagnol, et ce qui acheva de me le confirmer fut le récit qu'il me fit du massacre des Espagnols qui alloient à la découverte du Missouri dont j'avais entendu parler."

the West would be reached by way of internal exploration, can be easily seen by reference to his remarks, in the volumes of Charlevoix, in regard to his own cartographic work. He, namely, expresses his belief that Lake Superior is not more than three hundred leagues from the "Mer de l'Ouest," and adds that it is almost certain that there is "une suite de Lacs et de Rivieres, par lesquelles on peut communiquer de Lac Supérieur avec cette mer;" and the region is thus represented on Bellin's maps, as already mentioned.

A comparison of the various maps published about the middle of the last century shows how extremely vague were the notions of people in regard to the trend and position of the northwestern coast of North America. The map accompanying Hennepin's Travels arranges things so as to accord as completely as possible with the theory of reaching the Sea of the West with ease from the region of the Great Lakes. The coast north of 45° is made to trend to the eastward so as to bring it only a little over 20° in longitude west of the head of the Mississippi; "La terre de Jesso" and the upper part of the Gulf of California are in the same latitude, and only distant five degrees of longitude from each other. In conformity with the idea already alluded to as prevalent during the earlier part of the last century, that Asia could be easily reached from the Great Lakes of North America, the map accompanying Venegas's "Noticia de la California," bearing the legend "Formado sobre las Memorias mas recientes y exactas hasta el año de 1754" gives an independent existence to the Sea of the West, distinguishing it from the Mar del Sur or Pacific, calling it the "Mar o Bahia de el Vest," and representing it as a vast interior sea extending from the opening discovered by Aguilar east to within five degrees of the Lake of the Woods and north to the latitude of the central portion of Hudson's Bay. On Carver's map (1778) the legend "Western Sea" occupies a space only about ten degrees west of Winnepegosis, but no attempt is made to give its boundaries. This map, however, is simply an exact copy of Palairet's map, of thirteen years' earlier date, as already mentioned. The first discovery of the point at which the "River of the West" enters the Pacific was made by Heceta, on the 17th of August, 1775. This navigator did not, however, enter the mouth of the river, as he found the difficulties greater than he could overcome, nor did he positively ascertain that this was the mouth of a great river, although he surmised that it was. Hence this discovery of Heceta's must be looked upon as being a very unsatisfactory one. Meares, the English navigator, also failed to enter this river, although he was in the bay of the Columbia and gave to its northern headland the name which it still bears — that of Cape Disappointment. The disappointment was that there was no good harbor here, as he had been led by Heceta's account to expect. The entrance to the Columbia is indeed a dangerous and difficult one for sailing vessels, except under specially favorable conditions of wind and weather. Only steamers with skilful pilots can get in or out without liability to great delay or even serious danger, as is clearly shown by the great number of shipwrecks which have taken place on the bar of the Columbia, including that of the United States ship "Peacock" — one of the Wilkes Exploring Expedition vessels.

Vancouver was also unfortunate in missing the mouth of the great River of the West, although he noticed that the color of the water in the bay was that indicating the outlet of a river. The surf was breaking in a continuous line along the bar from highland to highland, north and south, and he did not dare incur the risk of crossing it, especially as he had no idea of the importance of the discovery he would have made had he been successful in getting over the bar.

These various navigators were either not acquainted with, or paid no attention to, the prevalent ideas of the French explorers from the East overland, with regard to the existence of a great river system, through which access could be had to the region to the northwest of the head of the Missouri. The relatively small importance of the rivers entering the Pacific from the western side of South and North America, as compared with the magnitude of those draining the Atlantic slope, may not unreasonably be supposed to have operated as a check on the search for great rivers on the Pacific coast. It is only fifty years since the character of the drainage of the part of the coast lying between the Colorado and the Columbia became known even in its rudest outlines; and the writer has within ten years purchased, from prominent map establishments in London and on the Continent, maps offered for sale as including all the newest investigations, in which this drainage was represented as it was believed to be before the explorations of Bonneville had revealed the existence of the "Great Basin." Fremont himself, when at Tlamath (Klamath) Lake, in 1843, expected soon to strike, in going south, the "famous Buenaventura River . . . forming, agreeably to the best maps in my [his] possession, a connected water line from the Rocky Mountains to the Pacific." He seems either to have been unacquainted with Bonneville's work, published six years earlier, or else to have ignored or misunderstood it.* And yet at that time Fremont was himself within the Great Basin, Klamath Lake, like many others in that vicinity, being without any drainage to the sea.

Mr. Robert Gray, commander of the ship

* Fremont could not have been entirely unacquainted with Bonneville's expedition, since he met with Joseph Walker, and mentions the fact that this renowned fur-hunter and explorer "was associated with Bonneville in his expedition to the Rocky Mountains." This is the only allusion made in any part of Fremont's reports to his predecessor and the unquestioned discoverer of the Great Basin. It would seem impossible that a work coming from the pen of so distinguished an author as Washington Irving could have been unknown to one who was about to undertake the exploration of the same region which Bonneville had visited a few years before. Bonneville was unfortunate in having Irving as the editor of his travels. The editing is badly done; but the maps mark an important step in the progress of geographical knowledge in this country.

"Columbia," which had sailed from Boston, Sept. 28, 1790, was more fortunate. On a second trial, he succeeded, May 11, 1792, in overcoming the dreaded obstacle, and entering the river, up the estuary of which he sailed about twenty miles. Lieutenant Broughton, of the "Chatham," one of Vancouver's expedition, shortly after this ascended the Columbia, as far as what he called Point Vancouver, about eighty miles from the mouth, where was afterwards Fort Vancouver, the chief trading-post of the Hudson's Bay Company on the Pacific. This is a little below the Cascades, and nearly opposite the entrance of the Willamette into the Columbia.

Captain Gray gave the name of his vessel to the river whose mouth he was undoubtedly the first white man to enter, but which was well known to the numerous natives along its course, and which, as we now know, was the great "River of the West" and the Oregon or Oregan of Carver. The name given by Gray to the river is the one which it now bears; but many years elapsed before this name became generally current. To see what names were in use at any particular time in the early history of the geography of this country we naturally look in the various editions of Morse's "American Gazetteer," or of the Geographies of the same author. In the Gazetteer, edition of 1797, we find only "River of the West" used as the name of what is now called the Columbia. In the edition of the "American Geography" of 1805, however, the new name "Columbia" makes its appearance in the text; but the accompanying map has only "Oregan." The river is said to "deserve notice, and to have been ascended in boats to more than eighty miles." Lewis and Clarke speak only of the Columbia River; and except that they mention its Indian name,* never once allude to its having at any time borne any other appellation in any part of its course, or to its identity with the "River of the West." Neither do they ever mention the name of Carver, Gray, or of any of the previous explorers of the region they visited, with the exception of that of Mr. Fidler, one of the geographers in the service of the Hudson's Bay Company.

But at the time the instructions for the Lewis and Clarke expedition were made out (1803) by Jefferson, it is evident that he

* Mentioned twice — once as "Shocatilcum" and again as "Chockalilum."

did not consider that it was known whether the Oregon and the Columbia were one and the same river; for he expressly directs them to find out whether "the Columbia, Oregan, Colorado, or any other river may offer the most direct and practicable water-communication across the continent for the purposes of commerce."

The explorations of Vancouver on the northwestern coast of North America had just about this time put an end to the prevailing condition of uncertainty with regard to a possible unbroken navigable communication between the Atlantic and the Pacific; but it remained to ascertain how far rivers could be made to take the place of the ocean, and what difficulties would have to be surmounted in order to get from navigable water on the Atlantic slope to the same on the Pacific side of the continent.

The Arrowsmith maps are those which contain the geographical information collected by the Hudson's Bay Company, so far as they saw fit to allow it to be given to the world. The Arrowsmith map of North America, which Lewis and Clarke had with them, or used when writing up their notes for publication, is evidently that of 1795, perhaps "with corrections to 1802." On this map nothing is given in regard to the region at the head of the Missouri, or of that traversed by any of the head-waters of the Columbia, which could have been of any use to explorers of that region.*

How vague the knowledge of the geography of this part of the country was at the end of the eighteenth century, and even during the earlier years of the nineteenth, is shown by the fact that Mackenzie, when he crossed over the divide between Peace River and a river which, as the natives assured him, ran into the salt water, did not know at all what river he had reached. In his journal he carefully avoided giving it a name, calling it only "the great river." It was in fact the Fraser; and learning from the Indians how much it was obstructed by rapids, he, after having descended it for a short distance, retraced his steps, and made his way to the Pacific by land, crossing some of the southern affluents

* The "remarkable mountain called the Tooth" (*vide* Lewis and Clarke, Am. ed. vol. i. p. 253), laid down by Arrowsmith from Mr. Fidler's observations, it seems quite impossible to recognize or locate. The same may be said of Mr. Fidler's other names given to various points in the Rocky Mountains, no one of which appears to have become current.

of Dean's River, and reaching the head of the Bentinck Arms, near what is now called New Aberdeen. This was in 1793, and the narrative of his journey was not published until some years later, — namely, in 1801. By this time the name given to the Columbia River by Gray in 1789 had become known, and Mackenzie had heard of it; for in the general remarks at the close of his volume on the discovery of a passage from the Atlantic to the Pacific, he speaks of the Columbia, calling it "the Tacoutche or Columbia," but not in such a manner as to render it certain whether he did or did not consider the "great river" of his journal to be the Tacoutche. Arrowsmith, however, did so consider it; for his map, published about that time, part of which is included in Mackenzie's volume, shows the river which this explorer reached as connecting by a dotted line with the Columbia, which is laid down from Broughton's Survey, and the whole is named the "Tacoutche Tesse (*Tesse = river* in Chipewyan) or Columbia River."

Humboldt as late as 1811 — the date of the publication of his great work on New Spain — was still very much in the dark as to the identity of the Tacoutche with the Columbia, which he calls the "Tacoutche Tesse ou Orégan de Mackenzie." It should be "of Arrowsmith;" for Mackenzie never used the name "Oregan," or "Oregon," and he evidently depended on Arrowsmith entirely for his general cartographic ideas outside of the field of his own especial explorations. Humboldt at that time was decidedly inclined to consider the Oregon as being distinct from the Columbia, and as perhaps emptying into Great Salt Lake ("un des grands lacs salés que, d'après les renseignemens donnés par le père Escalante, j'ai figuré sur ma carte de Mexique sous les 39° et 41° de latitude"). Humboldt calls attention to a curious blunder of Malte-Brun's, who in his geography had seemed to recognize the name "Oregan" on a map of Mexico published by Antonio Alzate, on which in regard to the Colorado River it is said, "cuyo *origen* se ignora" (whose source is unknown).* Malte-Brun at this time agreed with Humboldt in inclining to the belief that

* It is not a little curious that Humboldt makes, in his "Ansichten der Natur," a slip almost as amusing as that of Malte-Brun, which he twice records, once in his "New Spain" and again in the "Ansichten der Natur." In the last-named work he remarks that Malte-Brun thought he recognized the name "Oregon" in the *ignora* of "aun se ignora." Of course it should have been, as above, the *origen* of "cuyo origen se ignora."

the Oregon was distinct from the Columbia; but the first-named geographer considered it more likely that the Oregon emptied into the Gulf of California, while Humboldt thought that this would be ascribing to that river a very improbable length, and therefore preferred to adopt the view that it was a river belonging to the Great Basin System, as, with our present knowledge of the region, it would be proper to say. It may seem strange that so much ignorance in regard to the topography of the Northwest should have prevailed as late as 1811; but it must be remembered that it was not until several years after the completion of Lewis and Clarke's explorations that the results of their memorable journey were given to the world. The very eccentric course of the Columbia proper, and the enormous distance between its head and that of its principal southern tributary — the Snake — are sufficient reasons why the hydrography of this region should have offered a puzzling problem to the early explorers; and it was not until about 1864 that the difficulties were all finally cleared up, although Fraser and Stuart descended the river named for the first of these explorers as early as 1808. The name "Tacoutche-Tesse," as that of the Fraser, is seen in maps published in England certainly as late as 1832, and perhaps much later.

From the time of the naming of the River of the West by Gray, this western region rapidly increased in importance, and the names of tribes, places, and rivers began to become more or less familiar to Europeans, and to the inhabitants of what was then the United States. The exploration of Lewis and Clarke brought the region at once into notice, and the vague way in which "Louisiana" had been described in the purchase of that vast and not distinctly limited region made it not unlikely that the claims of the United States would, sooner or later, be extended over it.

Although the name "Oregon" nowhere occurs in Lewis and Clarke's report on their expedition, yet this name continued in use, and gradually became a familiar word in the United States as well as in Europe. The Columbia River was called the Oregon, as well as the Columbia, down almost to the present day. Twiss says, writing in 1846, "The Great River of the West is best known in Europe by the name of Oregon." The quotation given at the head of this article, from a poem published in 1821, shows that the name was in this country also a familiar one at this time.

Bryant, in the original edition of "Thanatopsis," spells the name "Oregan," and the same spelling is preserved in the authorized edition of 1847. In later editions, and especially in that of Parke Godwin of the collected works of this author (1883), the name is spelled "Oregon." As a good illustration of the uncertainty in the use of the name, it may be mentioned that Flint, in his "History and Geography of the Mississippi Valley" (1832), twice calls this river the Oregon, on the very same page in which he says that the name of Columbia was given to it by Captain Gray. Indeed, it was not until the limits of the State of Oregon became fixed, and it had been received into the Union, that Columbia began to be the generally accepted name of the river. The State and the river might, however, easily have both retained the name of Oregon, as there are thirteen States which have the same designation as the principal rivers forming portions of their boundaries, or by which they are traversed (Mississippi, Missouri, Ohio, Illinois, Wisconsin, Iowa, Tennessee, Arkansas, Minnesota, Kansas, Connecticut, Delaware, Alabama). The popularity of the name "Columbia," and the desire to perpetuate the memory of the fact on which so much seemed to turn in the settlement of the boundary question with England — namely, that this river was first entered by a citizen of the United States — finally inclined the scale in favor of Columbia, especially after the name "Oregon" became definitively fixed as the name of a State.

While "Oregon" continued in common use as the name of the Columbia River, the same name was universally applied from a very early period to a region of indefinite extent bordering the Pacific Ocean. The area to which this name was generally given was that lying between the possessions of Mexico on the south, and of Russia on the north, on the Pacific coast; and it was understood that Oregon extended eastward from the coast as far as the crest of the Rocky Mountains. As soon as Oregon became a State, and the name of Washington was given to the region bordering the Pacific north of the Columbia River, the former name acquired a definite meaning, and at the present time by Oregon only the State of that name is meant.

We come next to a consideration of the meaning of the word "Oregon," first appearing in print in Carver's Travels, but of the origin of which he gives no hint. As before remarked, the word in question has been a stumbling-block ever since Carver's day. No one has ever professed to have discovered its derivation or meaning. Various — in fact, numerous — suggestions have been made by writers; but no one has been insisted on, or attempted to be sustained by evidence. In fact, the condition of things in regard to the meaning of this name remains, up to the present time very much as indicated by Greenhow: "As to the name 'Oregon,' or the authority for its use, the traveller [Carver] is silent; and nothing has been learned from any other source, though much labor has been expended on attempts to discover its meaning and derivation. It was most probably invented by Carver."

It has never been seriously claimed that Oregon was an Indian name, although Mr. Twiss does say, "The native name, however, will not totally perish in the United States, for it has been embalmed in the beautiful verse of Bryant." Neither is it French nor English. The only language left from which it possibly could come is Spanish; and here, too, it has been repeatedly sought, but apparently never by any one acquainted with that language and at the same time with the names of the Indian tribes inhabiting the region drained by the Columbia. The nearest approach made to the real signification of the word "Oregon" was that of Lieutenant Symons in his official Report on the Columbia River (p. 86), who has the following: "Although it does not seem possible to determine with certainty the origin of the word 'Oregon,' it does not seem at all probable that it is a meaningless word invented or coined by Carver. It has been claimed, and not without some reason, that it is from the Spanish word *Oregano*, the wild marjoram (*Origanum Vulgare*, L.), found growing in abundance along the coasts. It may also be from the Spanish word *oreja*, ' the ear,' or some of its derivatives, as *orejon*, or *orejones*, signifying ' dried fruits ; ' and in the familiar language of Spain signifies '*dogs*'-*ears*,' an '*ear-pulling*,' etc. A derivative word, *orejera*, signifies 'a sort of ear-ring worn by Indians.'" It seems strange that Lieutenant Symons, having got so near the origin of the word in question, did not advance farther in the same direction.

The name "Oregon" is unquestionably the Spanish word *Orejon*, as we will now proceed to show.

In the first place, it may be mentioned that on the Columbia River this word was admitted to be of Spanish origin. Alexander Ross, who was one of the Astor Expedition, and who published a book entitled "Adventures of the First Settlers on the Oregon or Columbia River," expressly calls this river the "Oregon of the Spaniards." Hence, if we can find a plausible or reasonable derivation of the word from the Spanish, we shall have no difficulty in accepting it.

It may seem to some that the evidence of the presence of the Spaniards on the various branches of the Columbia was not so clear as to warrant the belief that they gave a name to this river. We may here recall what has been already mentioned with regard to the expectations of the French explorers, that, in case they attempted to reach the ocean, they would be brought in contact with the Spaniards, of whom and of whose doings reports were constantly being brought to the French posts by the Indians. The abundance of horses along the Columbia and to the north of that river, and the skill of the natives in the use of that animal, are already strong proofs of the presence of the Spanish race in that region. The Indians not only had an abundance of horses, but they knew how to take care of them, and even had learned to use the lasso with dexterity. But Lewis and Clarke speak repeatedly of the near neighborhood of Spanish colonies from whom the Indians stole or purchased horses, and supplied themselves with bridles and stirrups. They do not, however, locate these colonies with any degree of exactitude. In the case of the Shoshones it is said that they could reach the Spanish settlements in ten days " by way of the Yellowstone;" but whether this means by going up or going down that river is not certain, since at the time of Lewis and Clarke's expedition the Shoshones lived alternately on the Columbia and the Missouri sides of the Rocky Mountain divide. The evidence, therefore, that the Spaniards during the last century were more or less spread over the Columbia basin is positive; and there is no reason whatever why they should not have had a name for that river, or why this name should not have been known to people of other nationalities over a wide area.

The question arises, then, What does *Orejon* mean, and how is it applicable to this river? *Orejon* is the regularly formed augmentative from *oreja*, "ear"— *orejon*, "big ear." This is the original meaning; and if it is not found with that meaning in dictionaries of the present time, this only shows that, like many other words, it has lost in part its original signification. An *orejon* at present is a slice, or "big ear," of a peach or some other fruit cut off and dried in the sun. The ear-shape of the piece thus prepared is sufficiently suggestive of the reason why it came to be thus named. If, however, we look in a Spanish dictionary two or three hundred years old, we find "*Orejon*, one that hath large eares." (Minsheu's Dictionarie in Spanish and English. London, 1599.)

Whence comes it that big ears have anything to do with the river called the "Big-Ear River," or, as it undoubtedly was in the original Spanish, "Rio de los Orejones," or Oregones, the River of the Big-Ears?

The Big-Ears are the Indians called by the Spanish explorers and traders in the region drained by the Columbia River the "Orejones," a word which would be more likely to be written by English-speaking travellers with a *g* than with a *j*, the first-named letter more nearly representing the sound of the Spanish *j*. The reason for the giving of this name will be easily understood in reading the following extract from Carver's book: —

"The young Indians, who are desirous of excelling their companions in finery, slit the outward rim of both their ears; at the same time they take care not to separate them entirely, but leave the flesh thus cut still untouched at both extremities. Around this spongy substance from the upper to the lower part, they twist brass wire, till the weight draws the amputated rim into a bow of five or six inches' diameter, and drags it almost down to the shoulder. This decoration is esteemed to be excessively gay and becoming."

Carver, who was sufficiently near the region inhabited by the people thus adorning themselves, to have heard of this custom, also heard at the same time of the name which the Spaniards had given to the river in the vicinity of which these Big-Ears lived. To the river called by the Spaniards the Rio de los Orejones or Oregones, he would naturally give in English the name "Oregon," although he perhaps did not understand its meaning or connect it with the method of aural decoration which he so carefully describes. At all events, he makes no mention in his book of any such connection. If he had been acquainted with the Spanish language, he could hardly have failed to recognize the origin of the name.

The name of "Oregons," "Orejones," or "Oregones" is not one known among the numerous appellations given to the Indian tribes of North America; and this might possibly lead to a doubt whether there were not some mistake about the fitness of the name as applied to a tribe or persons decorating themselves in the way described by Carver. Fortunately for the theory of the origin of the name "Oregon" here advocated, we are able to find, in another region colonized by the Spaniards, this very name actually in use at the present time as designating a tribe of Indians decorating themselves, if not in precisely the same way as that described by Carver, at least in a manner so nearly akin to it as to leave no doubt of the applicability of the name to the North American as well as to the South American "Big-Ears." The following is an extract from a paper entitled "Notice of Recent Journeys in the Interior of South America,"* by Alfred Simson: "The next tribe is that of the Oregones. . . . As their name implies, they have large ears, the lobes of which are bored and stretched until a block of wood up to an inch and a half diameter can be inserted. . . . The OREGON language is very agreeable to the ear," etc. Professor Raimondi, in 1869, visited this tribe, whose present residence is at or near

* Proc. Roy. Geog. Soc., vol. xxi. pp. 556–580.

Pebas, on the banks of the Amazon, within the limits of Brazil, but not far from the Peruvian boundary. This distinguished explorer of Peru describes the "Orejones," as he calls them, very much as does Mr. Simson, but says that the lobe of the ear is sometimes enlarged to such an extent as to touch the shoulder.* He adds that this "barbarous custom," as he calls it, is fast disappearing, since only the older people were seen with these peculiar appendages. It appears, indeed, that the Big-Ears were personages of importance in former times in Peru, since we read in Velasco's "History of the Kingdom of Quito," in regard to Huyana-Capac, as follows: " [They were] his best troops ... the flower of the nobility and of the army. Their distinctive mark was that they wore large golden rings in their ears, which fact, on account of the great size of these, caused them to be generally designated as the Orejones."

The name of "Orejones" or "Oregones" — the Oregons — as the appellative of a tribe of North American Indians seems not to have found its way into print; as distinguishing certain South American tribes having a fancy for this particular adornment of the aural appendages, it is, as has been seen above, still in use.

* Raimondi's El Peru, vol. i. p. 402.

The Big-Ears of North America seem first to have been made known to the world under the designation of "Ear bobs;" at least, this name is found on the Arrowsmith map of North America, edition of 1795, with corrections to 1811, given to the lake now called Arrow Lake, which is on that map designated as "Chatth-noo-nick or Ear bobs Lake," and the region adjacent to this lake has the name "Ear bobs" upon it, as indicating it as the residence of a tribe of Indians of that name. On a later Arrowsmith map the name "Ear bobs" has disappeared, Arrow Lake appearing with the name which it at present bears, while the lake which in the earlier Arrowsmith maps bore the name of "Kullespelm" appears as "Pend d'Oreilles or Kullispelm Lake," the river of which it is an enlargement being designated as "Clark, Pend d'Oreille, Flat Head, Kallispelm or Salish R." It is the central one of the three great branches which unite to form the Columbia River.

For many years the Big-Ears, or the Ear bobs of Arrowsmith, have been chiefly known by their French name, as indicated above; but sometimes this has been supplemented by a translation of the name into English; sometimes the Indian name of the tribe has been added; and, in a few cases, French, English, and Indian have been used by the same author, either at the same time or in different parts of his work. The number of variations in the spelling and translation from one language to another of the name of the lake in question, and of the tribe living adjacent to it, is large, as may be seen from an inspection of the following table, which, however, does not pretend to completeness, although it seems to include every possible change which could be rung on the name Pend' Oreilles. When the author has given a translation of the name, or its equivalent, either as applied to the lake or the tribe, whether in English, Indian, or both, it is so set down in the table.

AUTHOR	FRENCH	ENGLISH	INDIAN
Arrowsmith	. . . .	Ear bobs	Chatth-noo-nick
Bonneville	Pends Oreilles	Hanging ears	
Tardieu	. . . .	Ear Bobs	Cutsanim
Duflot de Mofras	Pend' Oreilles	. . . . . .	Kallespem
Domenech	. . . .	Ear-rings	
Grey	Pen d'Oreille		
Simpson and many others.	Pend' d'Oreilles	. . . . . .	Kullespelm
Parker	Pondera		
Brouillet	Ponderay		
De Smet	Pends-d'Oreilles	Ear rings	Kalispels
Many later writers.	Pend d'Oreilles		

The name in French has been supposed, as is seen above, to stand for either ear-rings or hanging ears, both being appropriate designations, — the one for the ornament itself, the other for the ear as thus adorned. The French would therefore correctly be, if written in full, in the former case, "Pendants d'Oreilles;" in the latter, "Pendantes Oreilles." The abbreviated form of the first of them should be, nearly as De Smet (a Belgian) has written it, Pend's d'Oreilles; or possibly as Governor Simpson and others have written it, Pend' d'Oreilles. It seems easier to take the word, in agreement with the French authors Bonneville and Duflot de Mofras, as meaning "Hanging Ears — Pendantes Oreilles" — and to abbreviate this into "Pend' Oreilles."* Twiss, Duflot de Mofras, and other educated men write it thus. The form now most usually adopted by mapmakers — "Pend d'Oreilles" — is certainly not correct, since there is no indication in the name as thus written that the syllable "Pend" is an abbreviation, as it must be, either of "Pendantes" or of "Pendants."

Although Carver so minutely and accurately

---

* This is the way the name is pronounced, however written.

describes the method of adorning the ear, which has, both in North and South America, been the origin of the name "Oregon," yet he does not locate the custom among any particular tribe or in any particular region. His description is simply a part of an account of the Indians in general. Neither does it appear that this practice was limited to any one tribe, although it is reasonable to infer that it was especially common among the tribe designated by the name of "Oregones," or "Pend' Oreilles."

Among the 150 portraits of North American Indians, mostly in full parade dress, included in M'Kenney and Hall's great illustrated work, quite a large proportion of the individuals portrayed have their ears more or less ornamented. In the majority of cases this ornamentation is effected by rings of beads or bugles which are inserted in perforations all along the internal border of the helix, and frequently hang down in festoons, so as almost or quite to conceal the ear. In several cases the lobe of the ear exhibits a large perforation, as if it had once been occupied by a block of wood, in the manner described by Simson in speaking of the South American Oregones, which block had been subsequently removed. There are two individuals among those depicted in M'Kenney and Hall's book which have the ear ornamented exactly in the manner described by Carver. In one of these two cases the ear-ring is much larger than in the other. The original picture from which this lithograph was copied is of life size, and is apparently very accurately painted; and as it is at present at the Peabody Museum in Cambridge, it is easy to give very nearly the exact dimensions of the elongated portion of the ear, around two thirds of which a thin plate of metal, or else a fine wire, apparently of brass, is bent or wound. The extended ear, as thus enclosed in its sheath or ring, hangs down so as just to touch the shoulder. The distance from the external auditory meatus across the longest diameter of the ring is five inches: a measurement at right angles to this gives about three inches as the transverse diameter of the oval. The name of this Big-Ear is given as Payta Kootha, signifying "Flying Clouds;" and he is said to be "a Shawanoe of the Chilicothe tribe, but born in the country of the Creeks, and in 1833 living west of the Mississippi." The other Indian with a similar aural appendage, but of lesser size, is also described as being a Shawanoe, or Shawnee — a migratory tribe which made its way gradually from the extreme southeast of the country into Virginia and Ohio, and still farther to the northwest. Nothing whatever is said in the text of M'Kenney's work as to the distribution of this peculiar method of ornamentation, but it was evidently not limited to any one tribe, and must certainly have been practised at about the same time by tribes separated from each other by a distance of at least 2,000 miles.

There have long been missions established among the Pend' Oreilles. Father de Smet, a Jesuit, has published two small volumes in regard to these missions among the Pend' Oreilles, the Cœurs d'Alênes (Pointed-Hearts), and the adjacent tribes. He is enthusiastic in his descriptions of the success of these missions, and of the good effects which they have had on the Indians, especially on that tribe which he calls his "dear Pends-d'Oreilles." De Smet's books date back about forty years. The later travellers have spoken well of these tribes, but of their precise condition at the present time the writer has no special knowledge. The Flat-Heads of this region still keep that name, although their heads are no longer flattened; and of the Pend' Oreilles a similar remark may be made, for it would appear that this method of decoration is no longer in vogue.

Since the above was written, the writer's attention has been called to an article by Mr. J. Hammond Trumbull on the origin and meaning of the name "Oregon," published in the "Magazine of American History,"* in which it is endeavored to be shown, — or, rather, it is distinctly stated, — that "the name is not Spanish." Every argument offered in support of this view has, in the preceding pages, been met, and, as the present writer believes, fully and satisfactorily answered. The name *was* recognized on the Columbia as being of Spanish origin; Spaniards *did* inhabit the region; there *is* a tribe of Indians dwelling on one of the main branches of the river to which the name is "peculiarly appropriate;" the word "orejon" *would* naturally be written "oregon" in English, as it has here been shown to be at the present time by persons writing about South America, because *g*

* Vol. iii. no. i. p. 36.

much more nearly renders the sound of the Spanish *j* than does the English *j*, and as proof of this we find that, in the early days of California, Spanish names having a *j* in them were frequently written, by those not familiar with that language, with a *g*, — as " Vallego " for " Vallejo," etc.*

Mr. Trumbull's theory of the origin of the word " Oregon " is : that " it comes from an Indian language, with which Carver had been for many years somewhat familiar, and it is the accurate *translation* into that language of the name by which, as Carver had reasons for believing, 'the Great River of the West' was designated by the tribes that lived near it. It is the Mohegan *wauregan*, the Abnaki *ourighen*, the Delaware *wuliexen*, the Massachusetts *wunnegan;* signifying in all dialects ' good,' ' fair,' ' fine.' " How a river on the western side of the continent came to be called by a Mohegan, Abnaki, or Delaware name, is thus explained by Mr. Trumbull. " The Indians through whose countries he [Carver]

---

* It is well known to those acquainted with the Spanish language that *g* and *j* were formerly interchangeable letters, although not so much so as *b* and *v*. An examination of works written in Spanish and published in America shows that this interchangeability of *g* and *j* still prevails in this country to a very considerable extent.

travelled all spoke either Sioux or Algonkin dialects. Neither of his interpreters (one was a Mohawk, the other a French Canadian) understood the Sioux, but the Algonkin designation of a 'Fair River'—*wauregan, ourighen,* or *alleghany,* according to local dialect— must have been well known to them and to Carver himself." That is to say, — as near as the present writer can make out, — Carver, before starting on his journey, had gathered information which led him to believe that the river now known as the Columbia was called, by the tribes residing upon it, by a name which meant "Fair River." What he learned among the Sioux convinced him that this was true, and he therefore gives us that name, not in the original, nor in the language of the Sioux from whom he got the information, but in some dialect of the Algonquin language; the reader can take his choice among these dialects, according as he thinks "Wauregan," "Ourighen," "Wuliexen," "Wunnegan," or "Alleghany," to be most like Oregon.

This theory involves the following improbabilities: — First, that Carver was sufficiently acquainted with recent English and French literature to have heard of the mythical story of Moncacht-apé and of his descent of what he called "La Belle Riviere" to the Sea of the West; Second, that Carver was so little acquainted with the geography of the Northwest as to believe that his Oregon could be reached in a few days of foot-travel from somewhere in the region inhabited by the Kansas Indians; Third, that the Sioux were sufficiently well acquainted with the languages spoken on the Columbia or Oregon to be able to give the native name of that river, and to explain its meaning; and Fourth, that Carver did not make known what that name was in the original, but translated it into a dialect of the Algonquin, because one of his *interpreters* (as Mr. Trumbull designates them, *servants* as Carver calls them) was a "Mohawk Canadian."

In regard to the first of these improbabilities, a few remarks may be added, since Mr. Trumbull has brought up the story of Moncacht-apé, and has seemed to think that it had something to do with Carver's name "Oregon." The individual in question, according to M. Le Page du Pratz,* was a "sage vieillard" belonging to the Yazoo tribe, whom he met near Natchez, at a time not specified. It is impossible to go into minute details in regard to the

* Histoire de la Louisiane, Paris 1758, vol. iii. pp. 83-140.

two journeys — one to the Atlantic and the other to the Pacific — of which the Yazoo Indian gave an account to the credulous Frenchman. It is sufficient to say that the narrative is a tissue of impossibilities from beginning to end. One or two extracts will be sufficient to show its character. Moncachtapé was on the Missouri, at a distance of a month's travel on foot from its mouth, when he found himself among the "Nation des Loutres." From here he was accompanied by a native of that tribe and his wife, who "se croyoit prête d'accoucher," and who, for some unknown and mysterious reason, wished to be confined on what, if the story be true, would be the other side of the Rocky Mountains. With these two companions the Yazoo Indian travelled nine "petites journées" up the Missouri, and then turned *north* and travelled five days more, at the end of which time — as he says — "Nous trouvâmes une Riviere d'une eau belle et claire; aussi la nomment-ils la Belle Riviere." This river he descended in a "pirogue," without difficulty, to the "Grande Eau," over which bearded people were in the habit of coming every year (Japanese, Du Pratz calls them) in ships, to get a yellow, ill-smelling wood, with which they were able to dye a beautiful yellow. After remaining some time here, and helping the natives fight a battle with the bearded dye-stuff collectors, Moncacht-apé returned home "par la même route qu'il avoit tenue en allant."

It does not appear that the Yazoo Indian's name of "La Belle Riviere" ever obtained a place upon any map, except on that of Du Pratz himself, on which it is represented as heading somewhere in the vicinity of the present town of Bismarck; and it is in the highest degree improbable that the story of Moncacht-apé should have been known to Carver while he was in America. He had no maps with him during his journey, and his cartographic work was limited to putting his route, as nearly as he could guess at it, on such maps as he found current in England when he went there to prepare and publish an account of his journey.

TOPOGRAPHICAL NOMENCLATURE.

I. INTRODUCTORY.

IT is proposed, in the present section of this little work, to discuss the origin and meaning of the names given in the United States to prominent topographical features of the earth's surface. In doing this, it will soon become evident to the reader that it would be impossible to limit our range to one country or one language. It will be seen that, owing to the vast extent of the territory embraced within the limits of this country, and to the manner in which portions of it have been occupied from time to time by races speaking different languages, names of natural objects or features of the landscape are current, in some sections of the country, which are not English — that is, which are not current in England except as they have been carried from the United States back to the land of the mother-tongue. We shall find also that there are words which are perfectly familiar to the people of one portion of this country, but which are quite unknown in other sections except through books. Furthermore, we shall find that the number of words used to designate the various natural features of the earth's surface is large — much larger, in fact, than would have been expected previous to making a special study of the subject. Indeed, so numerous are these words, that it cannot possibly be claimed that the list will be exhausted in the present attempt to bring them together. A beginning may, however, be made on the present occasion, and the subject taken up again for a fuller treatment at some future time.

Of all the terms which are mentioned in the following pages, there is not one which comes to us from any of the aboriginal or "Indian" tongues once spoken over the region now occupied by the United States. A considerable number of Indian words form all or part of various proper names, and have thus become quite familiar to us — as, for instance, "sipi," "minne," "squam," "kitchi," and many others; but no one of all these words has been generalized so as to have become applicable to any class or form of scenic feature. We do, it is true, to a very limited extent find it convenient to make a distant approach to such a generalization of certain names; as, for instance, if we should say — as has been said — that the Shoshone Falls are a smaller "Niagara," or that the Hetch-Hetchy Valley is almost a "Yosemite;" but this is not carried far enough to justify us in putting either Niagara or Yosemite in any dictionary other than one of proper names.

Setting aside, therefore, as seems necessary, all consideration of the aboriginal tongues, even a very elementary knowledge of the historical development of our country will suffice to make it clear that we shall have in the main to deal with three languages — English, French, and Spanish. Not that the United States are not occupied, and very extensively occupied, by people speaking other tongues than these; but with the exception of a few words which have come down from the early Dutch settlers, there is hardly a trace of nationalities, other than those mentioned, in the entire range of our topographical nomenclature. Among words which we must call English, because they are in familiar use in England, there are many which belong to the Celtic and Scandinavian families, some of which have found their way to this country, although many of them are unknown to us except through the reading of English books.

It is the features of the land surface of the globe which here particularly demand our attention; but as a preparation for that which is to follow, a few lines may be devoted to the consideration of the nomenclature of the water. The most important division of the earth's surface is into land and water; and the coast-line of a country is, for any region wholly or in part bounded by the ocean, that feature which first claims the attention of the investigator of its geography. The first step in geographical discovery was to establish the shore-lines of the continents, or the great land masses of the globe; and the next was to fix the position and determine the outlines of those smaller areas of land which, not being large enough to be called continents, receive the name of islands. Only in the case of Australia, with its three million square miles of land, is there doubt whether the designation of continent or island would be more appropriate.

The most general and most satisfactory division of the land is into two parts — the Old World and the New. Asia and Europe

(Eurasia) belong together, the line of separation between the two being purely an artificial one. Africa was not long since joined to Eurasia, but has only within the past few years been artificially separated from it. Hence the Old World is essentially one land mass, with very numerous islands attached to it, especially on its southeastern side, one of which, as before remarked, is almost or quite large enough to rank as a separate continent. The New World is, with the exception of its extreme northwestern corner, entirely and widely separated from the Old World. It is naturally subdivided into two portions, which are in connection with each other, and yet by so narrow an isthmus as to have led many to believe that an artificial separation of the two parts would be possible. Indeed, at the present time a large expenditure of money is being made for this purpose.

By the simple word **ocean** is meant the whole body of water which envelops and covers almost three quarters of the surface of the globe; but when the ocean is spoken of in a general way, without reference to any particular portion of it, it is often called the **sea**, and its edge the "sea-shore," but never "ocean-shore," although it is allowable to say "shores of the ocean." Shakspeare uses "sea" and "ocean" synonymously, but the former much more frequently than the latter. With the English poets in general the two words are synonymous; and both are used in close proximity to each other, according to the requirements of rhyme and metre. By physical geographers the ocean is subdivided into five areas, each of which is considered a separate ocean, although these divisions are largely artificial, the lines by which they are indicated being in no small part parallels and meridians. Seas, gulfs, bays, sounds, straits, coves, holes, harbors, etc. are the names of the minor subdivisions of the ocean, or of such portions of the water surface as are more or less completely "land-locked," or separated by capes, headlands, or sinuosities in the coast-line. The nomenclature of these subdivisions is in general simple and easily understood, and it is not proposed to enlarge on them in the present connection. It is with the names of the various portions of the land surface of the globe that we here have to do: the water will only be considered when its presence is necessarily connected with the land in the scope of the definitions under consideration.

There is one all-important feature of the earth's surface, from the point of view of topographical nomenclature, and this is *form*. To this everything else is subordinate. The landscape — and by "landscape" is meant the total impression made on the artistic or educated eye by such portion of the surface as is embraced within the field of vision — is a complex thing. Form is usually the prime factor in the impression produced; but this is not always the case. Besides form, there are color, and light and shade; and the resulting effect may vary greatly according as the landscape is seen under a more or less favorable illumination or at different seasons of the year. A region in the highest degree monotonous when every object is wrapped in a sombre rain-cloud may be transformed into beauty by the glow of a rising or a setting sun. Of all this, nomenclature takes but little heed. In the names given to the more level portions of the earth's surface, however, where form is wanting, there the character and distribution of the vegetation become all important, as will be seen farther on.

II. MOUNTAINS, PEAKS, AND SIERRAS.

THE surface of the land, when looked at from the most general point of view, consists of mountains, valleys, and plains. These are the most comprehensive terms which can be used in English for regions conspicuously elevated above the adjacent land, for depressions within such elevated regions, and for areas which preserve a certain uniformity of level, and over which absence of considerable elevations and depressions is the important topographical feature. This seems a very easily comprehended statement of a very simple fact; and yet, when we come to look more closely into the matter, we find great complexity in the forms in which mountains, valleys, and plains exhibit themselves in different regions, and a surprising — one might almost say bewildering — variety of names which are applied to these various forms. This is often true for regions inhabited by people speaking one and the same language;

for local peculiarities of the landscape and dialectic variations of the mother-tongue may give rise to names, some of which are current only within very circumscribed areas, but all of which essentially form a part of the language, and which for that reason must be studied. But in investigating a subject of this kind we are led, almost as a matter of necessity, to take a wider range, and include more than one language within the scope of our inquiry, because there are few important divisions of the earth's surface over which only one tongue is spoken, and fewer still in which there is not more or less mixture of various languages, offering words which are relics of former races of inhabitants, or which for various reasons have been borrowed from other countries, and whose study may lead to interesting historic results. For instance, no one could investigate the orography of the Alps in any detail without the aid of some knowledge of Latin, Italian, French, and German, as well as of various dialectic forms of these languages. For the Pyrenees we need both French and Spanish, since that chain is divided between nations speaking those tongues; and in France we find the Celtic element becoming of importance in the topographical nomenclature, while in Great Britain it is still more prominent, the composite character of the English language showing itself in the most marked degree in the wealth of names of the features of the landscape which we there find current.

While the present inquiry has especially to do with English words in use as topographical designations, we shall not hesitate to seek for light in the study of other languages, to which indeed we are naturally led by those circumstances connected with the former occupation of large portions of our present territory by people not having English for their mother-tongue, as has already been mentioned.

By the term "orography" is meant the investigation of the forms and structure of mountains and mountain-chains, and it needs but little orographic study to find out that a single entirely isolated mountain is something of comparatively rare occurrence in Nature. Almost without exception, every mountain belongs to a "system of mountains" — to a "group," "range," or "chain." Indeed, most of the great mountains of the world belong to great mountain-chains, and have around them other summits of similar character and of somewhat nearly the same elevation. This is due to the fact that mountains are the result of general causes, which have been active on a grand scale and through long periods of time, and not of such as were confined within narrow limits. The most striking exception to this general statement will be found in the fact that volcanic cones are sometimes quite isolated, and rise, in such isolation, high above the adjacent region. As a remarkable instance of this, the grand cones on the plateau south of the Colorado River may be mentioned, as well as those which extend in an east and west line across Mexico. But the Colorado volcanoes are almost near enough to each other to form a group; and those of Mexico, in spite of their isolation, may well enough be taken as belonging to one chain. Etna, however, rises in solitary grandeur; and Vesuvius, with Somma as a portion of the once united whole, towers high above the minor cones in its vicinity.

Mountains, then, as a rule, occur extended over elongated areas of the earth's surface, occupying regions where elevations, foldings, breaks, and protrusions of the stratified and unstratified masses of which the earth's crust is made up have taken place along lines which are sometimes of extraordinary length, and which are generally so grouped as to have the element of length greatly predominating over that of breadth. Hence we find that most mountains are grouped in such a way as to form what are called "chains" or "ranges," the two words being nearly synonymous. Several "ranges" make up a "system" of mountains. There is a very general tendency to designate by the term "chain" a succession of high points connected by lower ones in such a manner as to impress upon the mind the fact that, in spite of these differences of altitude, there is an essential unity in the mass thus designated. In accordance with this, we find the word "chain" (Lat. *catena*, Fr. *chaîne*, Sp. *cadena*, Ger. *Kette*, etc.) in common use where mountains are written about scientifically. Quite analogous to this use of the word "chain" is that of "cord" or "string," which, however, we do not have as topographical terms in English; but which, as such, are of frequent occurrence in Spanish, in the form of "cordon" and "cordillera," both derived from "cordel" (Lat. *chorda*), a cord, or rope — the one being defined in the dictionary of the Spanish Academy as "mountains stretching over a long distance;" the other as "a chain (*cadena*) of mountains." * This word "cordillera" is one of special interest to us, and we may well go somewhat into detail in regard to it, at the same time noticing another Spanish word which has become very familiar to the American ear — namely, " sierra."

" Cordillera " and " sierra " are nearly synonymous words in Spanish, the latter meaning primarily a " saw " (Lat. *serra*), and hence the jagged outline of a mountain-range, as projected against the sky, which we can call in English " serrated," although we cannot call the range with a serrated outline a " saw," as the Spanish do. The subordinate ranges which make up the system of the Pyrenees are usually called " sierras," and the number of these is large. The "Sierra Nevada" (Snowy Range) is an important and picturesque

* We say, in English, a " chain," but not a " string," of mountains. These verbal distinctions are somewhat delicate. We say a " string " of fish, and a " line " of trees, and also, though rarely, a " line " of mountains. We say, " his life hangs by a thread;" but "his life hangs by a string " would sound very queerly. So we say a " file " of men, but not of mountains. (See farther on, in connection with " defile.") And yet *chorda*, *linea*, and *filum* all have essentially the same meaning of "cord, string, or thread." " Cord " in the form of " cordel " (Fr. *cordelle*) is a word well known and formerly much used west of the Mississippi, meaning " to pull a boat up-stream by means of a rope or towing-line."

group of mountains in the south of Spain; and this name is familiar to us as that of what may with truth be called the grandest and most important single member of the Cordilleran system of North America. The Rocky Mountains are made up of a considerable number of more or less independent ranges, and these together form the eastern division of the Cordilleras. The Sierra Nevada and Cascade Range, with the associated and more or less closely connected Coast Ranges of California and Oregon, constitute the western border of the great complex of mountains. table-lands, and valleys which occupies the western third of that portion of the continent which belongs to the United States. But this nomenclature is not valid so far north as Alaska, which is a part of our territory, although separated from the main body of it by a wide interval. North of the 49th parallel the ranges — the system there beginning to be much diminished in breadth — are generally known collectively simply as the Rocky Mountains. The name "Sierra Nevada" was the first to appear on any map of North America, as designating either the whole or any part of the Cordilleran system.*

* See *ante*, p. 21.

The word "Cordilleras" is of special interest to us as being the most general designation of the system of ranges which borders the entire Pacific coast of both North and South America, forming the longest connected system of mountain chains in the world, its linear development being fully eight thousand miles. The South American division is called the "Cordilleras of South America," or the "Cordilleras of the Andes," or more generally simply "the Andes;" while the "Cordilleras of North America" are, for brevity and convenience, known as "the Cordilleras," that part of the country which they occupy being called the "Cordilleran Region." How this has come about has been already explained at length.*

Almost the entire mass of the South American Cordilleras is included in States having Spanish as their official language, although the aboriginal tongues are still current over various portions of the region. Moreover, nearly the whole of the North American Cordilleras was, not many years ago, also under Spanish control, that being the official language not only of Central America and Mexico (as it now is) but also of both Lower and

* See *ante*, pp. 15-27.

Upper California, while the settlements of that nation had spread themselves, to a considerable extent, through the Rocky Mountains, and over the great Central Plateau, to beyond the Columbia River.* Thus it has come about that Spanish names, derived from simple and homely words meaning "string" and "saw," have become familiar to us, English-speaking people, and are permanently fixed upon the grandest features of our topography.

Passing next to a consideration of the names which individual mountains or parts of mountain ranges have received, we find that their number is very considerable, and that the study of the sources from which they are derived opens an interesting field of investigation.

A mountain range is made up of a number of more or less distinctly marked elevations, and for each of these by far the most common designation is the term **mountain**, which may be applied to any high point or mass sufficiently elevated above the surrounding country or range of which it forms a part to be considered worthy of a distinctive appellation. **Mount** is simply an abbreviated form of "mountain;" and it seems to be largely a ques-

* See *ante*, pp. 31, 58, 59.

tion of euphony whether the point in question shall be called "mountain" or "mount," the usage being that if the word *mountain* is employed it follows the proper name, and if *mount* it precedes it. But there is a decided tendency to use *mountain* when a range or group of elevations is meant, and *mount* when a particular summit of this group is to be designated. Thus we have "White Mountains" and "Mount Washington;" "Green Mountains" and "Mount Mansfield;" "Adirondack Mountains" and "Mount Marcy," etc. In the term "Rocky Mountains" is included a large number of ranges; the entire *system*, as it might properly be called, occupying almost half a million of square miles. The subordinate divisions of this grand system are usually called *ranges;* as the "Sawatch Range," "San Juan Range," etc. One subdivision, the "Park Range" of Colorado, is — or was, a few years ago — frequently called "the Snowy Range," the exact equivalent of "Sierra Nevada" and nearly the same as "Himalaya." The name "Rocky Mountains" is frequently abbreviated in familiar language to "Rockies." Similarly, the "Appalachian System" is often called the "Appalachian Mountains," and also, more concisely, the "Appalachians." This latter designation may properly be employed by scientific writers; but the term "Rockies" would hardly be allowed in a geographical work.

As soon as we look for more specialized designations of single mountains, or for such names as are indicative of peculiarities of form or structure, we begin to find great variety in the nomenclature. Perhaps the most common term, next to "mount," for an individualized mountain is **peak**, which means simply a "point" or "pointed." But in the United States this designation is often applied to mountains which are not particularly remarkable for having pointed summits. This usage is rather common in the Rocky Mountains, where Long's, Pike's, Gray's, and other "peaks" might as well, or better, have been called "mounts." In the Appalachians there are a few summits designated as "peaks" — as, for instance, the Peaks of Otter, in Virginia — but, in general, this system of mountains is remarkably free from conspicuously pointed elevations.

The word "peak" is used both in French (*pic*) and Spanish (*pico*) very much as it is in English. As examples, may be mentioned:

"Pico de Urbion," in which heads the Duero River; "Picacho de la Valeta"—"picacho" being the augmentative of "pico." *The* pico, however, is the well-known island of that name in the Azores, on which is the "Pico Alto," the highest point of the group. There are several "picachos" in New Mexico and Arizona, and one or more in California. To the name "picacho" is occasionally added the word "peak"—a kind of reduplication of a name in two different languages, the origin of which is easily understood. A large number of high and generally pointed summits on the French side of the Pyrenees and in the French Alps bear the name of "pic;" for example, "Pic de Néthou," "Pic du Midi," "Pic du Frêne," "Pic du Pyramide," etc. This word also appears in the Pyrenees in the form of "pique," and its diminutive "piquette."

The form in which "peak" occurs in the Lake District of England is "pike," a name there given to any summit of a hill, but more generally to such as are peaked or pointed. "Scawfell Pike" (3,160 feet) is the highest point in England proper. It is the culminating summit of the group of elevations collectively known as "Scawfell," standing at the head of the Wastdale. Scawfell is an interesting name, concerning which information will be given farther on, under "scar," or "scaur," and "fell." "Pike o' Stickle," one of the two "Pikes" of Langdale, is another curious name. "Stickle" (A. S. *sticcel*, Ger. *Stachel*) means "a sharp point" — a word which we have only in the familiar name of a fish, the "stickleback," so called from the stickles or prickles on its back. In the name "Pike o' Stickle" we have a reduplication similar to that to which allusion has already been made.

The word "peak" appears, in the form of "pique," in a work published in London in 1679,* in which the summit of Mount Athos is called "the high Pique or Peer." The "Peak" of England, however, is not a *peak*, but a *plateau* — a picturesque region in Derbyshire, near Castleton, about five miles in length, from half a mile to two miles wide, and having an elevation of about 2,000 feet. The mountainous part of Derbyshire is frequently called the "Peak Country." It is — to quote the language of the author of "All about Derbyshire" — "a wide expanse of alternating moor and mountain, green val-

* The Present State of the Greek and Armenian Churches. By Paul Ricaut, Esq. London, 1679.

ley and glancing stream, limestone tor and forest ridge." This name, which may have been given because portions of the edge of this table-land present a " peaky " aspect when seen from a distance, is a very old one.* It appears in Heylyn's Cosmographia, the first edition of which was published in 1664.

A word of which the English form is **spit** is one of those most commonly used in the Alps to designate a sharp or pointed mountain summit. It appears in German as "Spitze," "Spitzli," and " Piz ; " in Italian, as "pizzo." As "spit" it is familiar to us in English as the name of the long, pointed utensil on which meat is roasted. As a topographical word it is limited to the sea-shore, with the meaning of a sandy, projecting, not very elevated point. Wedgwood says : " Root uncertain ; but it would seem reasonable to connect *spit* with *spike*, *spine*, and *spire*; all of these words contain the notion of a sharp point." " Piz " is the form of this word most used in the Grisons Alps, especially on the Engadine ; "pizzo," in the Italian Alps : thus, Spitzliberg, in Uri ; Piz Roseg and Piz Morteratsch, in the Engadine ;

* Skeat says of it: M. E. *pek;* 'the hul of the *pek*' = the hill of the Peak, in Derbyshire; Rob. of Glouc., p. 7. In the A. S. Chron. an. 924, the same district is called Peac-land = Peak-land.

Pizzo di Verona, in the Italian Tyrol; etc. The Latin "spica," a point, spear of grass, spike (Fr. *espi, épi*), appears in the augmentative form in Spanish as "espigon," with the same topographical meaning as "Spitze" in German.

Puy is a word in common use in Central France to designate any kind of a hill or mountain. In the volcanic region of Auvergne all the elevations are "puys." Although this word would seem to be allied to "pic," it is thought by etymologists that this is not the case. Jaubert, in his "Glossaire du Centre de la France," derives it from the Latin *podium*. The variety of forms in which this word presents itself in various parts of France is indeed bewildering. In the Eastern Pyrenees we find "pueche," "pech," "puch," "puig;" in the Landes, "poy" and "pouy." Another word, "peu," seemingly a variant of "puy," is in use in Southern France; thus, le Peu, a locality near Saint Sévère (Indre), and in various other places. "Pié" is still another form; thus, Pié Montaigu, in the Commune de Palais (Cher), and Pié de Bourges, near Clion (Indre).

From the Spanish "peña," a rock, a large number of topographical words have been formed, either as augmentatives or diminutives, or with various other terminations, so characteristic of the language, and sometimes so difficult of definition. These names are applied to mountains and mountainous regions, but evidently not with any particularly nice discrimination as to the peculiarities of form of the object to which they are given. Any rock or rocky point may be called in Spanish a "peña," which is perhaps itself a diminutive of "pen," a word having a wide range over Europe. "Peñon," the regularly formed augmentative of "peña," is a word very commonly used in Spanish to designate a high, rocky point. "Peñol" is the equivalent of "peñon;" and "peñoleria" means a district or region where such elevations are numerous. "Peñasco" is another augmentative, and is one very frequently met with in Spanish mountain nomenclature; "peñascoso" is the corresponding adjective form.* "Peña Blanca" is the name by which a certain prominent mass of quartz, on the summit of a hill, and forming part of the outcrop of the "Mother Lode," is known in the Sierra Nevada; it is also

---

* "Peñasco" is defined by Barcia as a "peña grande y elevada;" and "peñascoso," as a "sitio, lugar ó montaña donde hay muchos peñascos."

called the "Peñon Blanco" by some. "Pène," the French form of "peña," is also a word in use in the Pyrenees. "Pico de la Peñalara" is the name of a high point in the Sierra de la Guadarrama.

The following is a synoptical statement of the more important Spanish names for mountains and mountain ranges, with some remarks on their meaning supplementary to what has already been given : —

From the Latin **mons**. *Monte, montaña, montañuelo.* "Monte" sometimes means rather the forest than the mountain, because the forests, in many regions, are so closely limited to the mountains. In the Peruvian Andes "montaña" has a peculiar meaning. It is the densely forested region on the eastern slope of the range, the country being divided into three longitudinal belts, — the "Coast," "Sierra," and "Montaña," the "Sierra" being the region of the Andes proper.

Latin **chorda** (Sp. *cordel*). *Cordon, cordillera.* Already sufficiently explained.

Latin **serra**. *Sierra, serrata, serranía, serrano.* "Serrata" seems to be occasionally used as the equivalent of "sierra." A "serranía" is a region of "sierras," a mountainous district. "Serrano" is an inhabitant of a "serranía," a "mountaineer," with the meaning of a resident of a mountainous region, not an amateur climber of mountains. "Serre," "serrat," and "serrère" are various French forms of "sierra" used in the Pyrenees.

Latin **penna** (O. L. *pennus*, sharp, pointed). *Peña, peñon, peñol, peñoleria, peñaranda, peñalara, peñasco* (adj. *peñascoso*).

Celtic, related to Latin **spica**. *Pico, picacho, espigon;* the latter is an augmentative, with the meaning of "high, bold, sharply pointed hill," — much less frequently used, however, than "picacho," which is heard among the mountains wherever Spanish is spoken.

Latin **cirrus**, a lock, curl, and hence a crest of feathers or crest in general. *Cerro, cerrito.* A very common designation of a hill, especially if not very high, but rough and rocky.*

Latin **collum**. *Colina, collado.* These two words are, as generally used, synonymous, and are nearly equivalent to the French "colline," English "hill." "Collado," also written "collada," is sometimes used for pass.†

* Barcia says: "Cerro es la colina en que abundan riscos y piedras y cuyo terreno es escabroso."

† Barcia says of "collado": "Altura de tierra que no llega á ser monte." Caballero: "Sitio que va subiendo en cuesta, formando garganta en la montaña, por donde facilita su subida y bajada."

Latin **lumbus**. *Loma, lomita, lomería.* The use of words derived from "lumbus," meaning "loin," as a topographical designation, is akin to our use of "flank" for mountain side. A "loma" is an elongated gentle swell of the ground, a rounded, inconspicuous hill. A "lomería," a region of low, rounded hills; the "foot-hills," as the comparatively low undulating region along the western base of the Sierra is called in California.

Latin **ventus**, with "re." *Reventon, reventazon.* A topographical designation of somewhat uncertain application, which seems peculiar to the Pyrenees. The idea conveyed is that of a cliff or precipice which repels the wind, as the waves on the seashore are thrown back by the rocks against which they are dashed. Barcia defines "reventon" as "a mountain slope extremely steep, and to be climbed only with difficulty." But "el Reventon" is also the name of the pass which crosses the Sierra de Guadarrama ("the glorious ridge," as Ford * calls it) to the north of the route from Madrid to the Escorial.

Latin **quadro**, quadratum. *Esquerra,* spelled also *ezquerra.* A name given to mountains

---

* Ford's Handbook for Spain, 5th edition, London, 1878, p. 89.

terminating in square or tabular forms, or squarely cut at one end. The modifications of this word in French are numerous ; for example, " queyre," " caire," " quairat," etc. A part of the French Alps where this form prevails is known by the name of " queyras," and the valley or gorge at the base of Monte Viso is called " Val de Queyras."

It would be hardly worth while to attempt to enumerate all the names given to mountains in the French Pyrenees. A number have been already given, and it will be sufficient to add what Ramond says in regard to the mountain nomenclature of that region: " L'idiome des Pyrénées a bien d'autres richesses ; *Pique* et *Piquette, Tuque* et *Tuquet, Roque, Poey* ou *Pouy, Cau* qui se prononce *Caou, Serre* et *Sarrat, Vigne, Hèche, Soum, Coste, Pêne, Mount,* et vingt autres mots barbares et mal sonnans, échappés du celte et du latin, voilà autant d'appellations spécifiques qui modifient l'idée générale de sommet, au gré des circonstances accessoires de la forme ou de la nature de chacun." *

Pen and **ben** are Celtic words, meaning " mountain," " highland," or " headland ;" and these names are applied to a large number of

* Ramond, Voyages au Mont Perdu, Paris, 1801, p. 255.

localities, — either mountains, headlands, or something analogous in various parts of Great Britain, especially in Wales, Northern England, and Scotland. The word "peña" and its derivatives, so frequently heard in Spain, as already noticed, are evidently allied to the Celtic "pen" and "ben." The word "pen" is occasionally seen in poetry, as in the following quotation from W. Crowe: —

". . . save only where the head
Of Pillesdon rises, Pillesdon's lofty Pen."

Band is another word of Celtic origin (Welsh, *bant;* Gael. *beann*, a hill). It is a name quite commonly given to the summits of not very conspicuous hills in the Lake District: for example, Swirl Band, near Coniston; Taylor's Gill Band, and Randerson Band, Borrowdale. This word appears to be another form of "ben" and "pen."

There are certain words which are in current use throughout the Western United States, and which on that account are familiarly known all over the country, either by personal acquaintance with the objects to which they are applied or through books, but which are not much used as topographical designations in the Eastern States. The most important of these words are "bluff" and "butte."

The etymological relations of the word **bluff** do not seem to have been clearly indicated. Wedgwood connects it with the Dutch *blaf*, which he defines as "planus, æquus et amplus, superficie plana, non rotunda," and thinks the word derived, in the first instance, from the sound of something falling flat on the ground. Be this as it may, "bluff" is the term applied everywhere in the Mississippi Valley, and to a very considerable extent as far west as the Pacific, to the steeply inclined sides of the river valleys. Throughout the prairie region the ascent is made from the river bottom to the rolling prairie by a sudden, sharp rise, the difference of elevation between bottom and upland varying in different regions from a few feet to several hundred. These abrupt rises, which are sometimes rocky and precipitous, and which might properly be designated as "cliffs," are almost universally known throughout the Mississippi Valley as "bluffs." In a portion of Wisconsin, however, there are several highly picturesque, isolated masses of rock quite castellated in form, which bear no other name than that of "bluff" among the people living near them.

In Holderness, Yorkshire, some detached hills are called **barfs**; one in the Vale of

York is also thus denominated. There seems to be a connection between "bluff" and "barf," and of both with "blaf;" but the matter is not clear. "Bluff" as a topographical designation is of comparatively modern origin. Cook, in his Voyages, speaks of a "bluff point," meaning thereby a "steep headland."

The French word **butte** is one often heard throughout the Cordilleran Region and on the Plains. As used in France, it is almost exactly the equivalent of our word "knoll," meaning a gently swelling eminence, or inconspicuous rounded hill. It is a word of rather uncertain etymological relations, but appears to be allied to (French) *but* and *bout.* It was introduced into the United States by the French trappers and *employés* of the Hudson's Bay Company, and has gradually come to be used as the designation of mountains of all degrees of grandeur, up to Shasta itself,— 14,440 feet high, and second only to Rainier among the great volcanic cones of the Sierra Nevada and Cascade Range. This grand mountain was "in the early days" of California most generally known as "Shasta Butte;" but this name is gradually giving way, and that of "Mount Shasta" taking its place. There are a number of conspicuous, more or less isolated mountains in the northern portion of the Cordilleras which are still called " buttes," and which are likely to be so denominated for a long time to come : for example, Medicine Butte (called also " Pill Hill"), Pilot Butte, Church Buttes — all in Wyoming ; as well as many in Colorado and the adjacent Territories. A high, craggy mass of rock, forming the crest of the Sierra in Yuba County, California, is known as the " Downieville Buttes ; " and the remarkable, isolated, and lofty volcanic range in the Sacramento Valley is called the " Marysville Buttes."

Knob is the favorite — indeed, almost exclusive — designation of any more or less isolated hill or mountain throughout the Southwest, and especially in Tennessee and Kentucky. There are various " Pilot Knobs ; " but the most famous one is that of Missouri, which is so largely made up of iron ore, and which is not far distant from the still more celebrated "Iron Mountain." There is an occasional "knob" in the Eastern United States, both in the White Mountains and in the Catskills ; but the topographical use of this word is decidedly a Southwestern peculiarity. "Knob" and "nab" seem to be the same words, the latter form being in common use in Northern England, and especially in the Lake District: for example, Nab Scar, Rydal; Nab Crag, Patterdale; etc. In Northeastern Yorkshire the abrupt hill edges are called "nabs."

Knock is another word belonging with "knob" and "nab." There are several "knocks" in the Lake District of England and in Scotland; as Knockmurton, Knock Pike, Knock Craig, etc. This word has been carried as far away from home as Australia and the adjacent islands. "Knocklofty" is the name of a "respectable eminence" near Hobart, Tasmania.

The words just enumerated run, with many variants, through the languages of Northern Europe. Thus, Gael. *cnap*, to strike or beat, and hence, as a substantive, that which is produced by beating, namely, a lump, or boss; and, going a little farther in the same direction, a hillock, or hill, which is also the meaning of *cnac* and *cnoc* in the same language. The Welsh is *cnwpa*, a knob or club, and *cnwc*, a lump or bunch, etc.*

Again, **knoll** and **knot** are closely allied to the words just mentioned. The former is in common use, both in this country and in

* See Wedgwood, under "Knob."

England, as designating a low, rounded hill. "It was a rocky knoll, that rose forty feet above the surface of the water," Cooper says in describing the locality which gave the name to the story called "Wyandotte, or the Hutted Knoll." "Knot" seems peculiar to the Lake District of England, where some of the hills of bare rock are called by this name; for example, Hard Knot, Farleton Knot, Arnside Knot, etc.

Dodd is another name for mountain, current in the Lake District, but not heard in the United States. This word is another of the same class as "knob" and "nab." It is defined in the glossary attached to Black's Guide as being a "hill with a blunt summit attached to another hill." In Frisian — a language closely allied to English — "dodd" means a "bunch." "To dod," in English, is "to cut off an excrescence." "Toddi," in Icelandic, is "a fragment or piece cut off." "Todi," like "pen," has a wide range. The word appears in the name of the grand peak of Glarus, the "Todi," or "Tödi;" and perhaps "dolde" in Doldenhorn, in the Bernese Alps, may be the same word. The following are examples of mountains in the Lake District bearing the name of "dodd:" Skiddaw

Dodd; Hartsop Dodd, Kirkstone; Dod Fell, near Hawes.

Mound, which, as usually applied both in England and America, means an artificial eminence of no great altitude, is the name given in Wisconsin, Illinois, and Iowa to those isolated flat-topped hills which occur in and near the Lead Region, and which rise a few hundred feet above the adjacent nearly level country. The West Blue Mound is about 500 feet above its base; others, as Platte, Sinsinnewa, Sherald's, Waddell's, etc., are from 200 to 300 feet in height. They are conspicuous objects in the vicinity of Galena and Dubuque, and they are all capped with the harder Niagara limestone, while their lower, more gently sloping portions are made up of the soft, shaly, and easily disintegrated strata of the Hudson River group, which lie between the Niagara and the lead-bearing limestones — or, more properly, dolomites, the latter being, in general, throughout the region the rock which occupies the surface of the country.

The word **cobble** as the name of a hill of moderate elevation is heard occasionally in this country, but, so far as known to the present writer, not in England. There are at least two "cobbles" in the Adirondacks, and one or more in Berkshire, Mass.; and there is also a Cobblekill (mountain rivulet) in the Catskills. "To cob," in English, is "to beat or strike" — a word in common use among miners, the breaking of masses of ore into small lumps being called "cobbing." "Cobble," as meaning a rounded fragment of rock, not so small as a "pebble" and not so large as a "bowlder," is also a word of every-day occurrence, both in England and America. "Cobble," as meaning a rounded stone, seems to be a diminutive of "cob," a lump: "cobble," as designating a mountain, is a word not so simple in its etymological relations. "Koble" is an old German word having the meaning of "rock" and "mountain," and "Kofel" has the same signification in the Bavarian Alps, Tyrol, and Carinthia. Many of the more prominent peaks in Tyrol are thus designated, the word being frequently written "kofl." Large stones are also called by the same name, and "köfeln" is "to throw stones."

Again, "Kopf" (head) is a common German designation of a prominent, rounded mountain summit, especially in the Bavarian and Austrian Alps — thus, Rosskopf, Ochsenkopf, Schwarzkopf, Adlerkopf, etc.; and the corresponding English word is by no means a rare name for mountain in English-speaking countries. Thus, we have a "Blackhead" in the Catskills, a "Bullshead" in the Southern Appalachians, a "Doublehead" in the White Mountains, etc. "Cob," "cop," "kop," "kopf" are all one and the same word originally, with the meaning of "rounded lump, or head," the association or meaning being evident and natural, as we see from the fact that large cobbles are, in this country, frequently called "nigger-heads."

The word **kogel** as a mountain name, frequently heard in the Austrian Alps, either as "kogel" or "kogl," — for example, Kreuzkogl, Ankogl, Graukogel, near Gastein, etc., — is apparently related to "kegel" (cone), a name naturally given to conical summits, and especially to those of volcanoes, which are often as regular in form as they are graceful in their proportions.

There are various names for mountains, the origin of which is quite obvious. They are given as indicating a resemblance to some familiar object. Some of these designations run through several languages; others are limited in their range. The following may be mentioned: —

The **crest** of a mountain or of a mountain range is a familiar expression. By it we mean the culminating ridge or "backbone" of the elevated mass. The same term is used in French (*crête*), in Italian (*crista*), etc. "Crest" is from the Latin *crista*, a word which is allied to *crinis*, hair; hence, something which grows on the top of the head like the "comb" or "crest" of the cock. The German equivalent for mountain crest is "Kamm," an old word appearing in the different dialectic divisions of the language as "Kam," "Kamp," "Chamb," etc., and having originally the meaning of "tooth" or "toothed implement." Thus we have, so to speak, got back to "sierra," the use of this word and of "Kamm" having exactly the same underlying idea. We have not in the United States the word "kam" or "kamm" as a topographical designation, but it is current in the Lake District of England: for example, Catstycam, or Catchedecam, Helvellyn; Rossthwaite Cam; Cam Fell, near Hawes; etc. It is not a little curious that the word "comb" also appears in that district as the name of a mountain. "Black Comb," also spelled "Combe," is a summit overlooking the Vale of the Duddon, and commanding a very extensive view, as thus poetically indicated : —

"Far from the summit of Black Comb (dread name
Derived from clouds and storm!)"
Wordsworth.

"Close by the sea, lone sentinel,
Black Comb his forward station keeps."
C. Farish.

Thus, "comb" and "combe" or "coom" are seen to be two words quite different in origin and meaning, although not unfrequently the two are spelled alike. The latter is a hollow in the mountain side; the former, the mountain crest.*

The **spurs** of a mountain are the subordinate ridges which extend themselves from the crest, "like ribs from the vertebral column" (Bonney). It is between the spurs that the water derived from the rain or from melting snow makes its way downward; and it is chiefly by the erosive action of this water, assisted by ice where the range is high enough to be glaciated, that the gorges, ravines, and valleys have been eaten out, leaving the spurs on either hand, as witnesses of the power of the erosive agencies.

The peculiar form of many mountain summits has led to a wide-spread use of the word

---
* See farther on, p. 167.

dome. Granitic masses frequently assume this shape, and with such perfection of outline that the use of this designation seems entirely natural. There are several mountains called "domes" in the Appalachians; but forms of this kind are most abundantly and characteristically displayed in the Sierra Nevada, especially in the neighborhood of Mount Whitney, and near the Yosemite Valley, where the concentric structure of the granite is developed on a grand scale, giving rise to dome-shaped masses of great regularity of form and of immense size. One in particular in the cañon of the North Fork of the San Joaquin, rising to a height of 1,800 feet above the river, looks like the top of a gigantic balloon struggling to get loose from the rock in which it is imprisoned.* A similar use of the word "dome" is not uncommon in the Alps, although there are no mountain summits in that range so perfect in their dome shape as are many in the Sierra Nevada.

The rounded summits of the Vosges Mountains are also called in French "ballons," and in German "Belchen" or "Bölchen." "Ballon," of course, is the equivalent of "balloon," and is etymologically connected with *balle*, Eng.

* Geology of California, vol. i. p. 401.

ball. There are six summits in the Vosges which are called "Belchen;" and the most famous of them is the Gebweiler Belchen (Fr. Ballon de Soultr), the culminating point of the range, commanding a magnificent view of the Jura, a part of the Alps being also visible in clear weather. "Boll" in German means "rounded, swollen;" and "Bolle" is "bud." A "bölliger Berg" is a mountain with a swollen or rounded top; hence the name "Belchen" or "Bölchen," used not only in Vosges, but to a limited extent in the Jura and Black Forest.

Objects more prosaic and familiar than domes give names to mountains — names which are sometimes limited in their range, but often occurring over wide areas and running through various dialects or even languages. Thus, **horn** is the most common designation of the highest peaks in the German Alps, and more especially in the Bernese Oberland; for example, Finsteraarhorn, Schreckhorn, Matterhorn, etc. The French use the equivalent word "corne" to a limited extent; and "corno" plays the same part in mountain nomenclature in Southern Italy. This use of the term "horn" seems quite unknown to English-speaking people, nor does it appear to have extended itself into Spain.

Dent, or tooth, takes the place of "horn" in the French Alps, and is one of the most common designations of high, more or less isolated peaks in that part of the chain; for example, Dent du Midi, Dent Blanche, Dents de Bertol, etc. The similarity of a sharp mountain peak to a tooth seems not to have impressed itself on the English mind, and there are few if any summits in England or in this country which are thus designated. The first name, however, given to any individual mountain peak in the whole Cordilleran system, within the limits of the United States, was that of "Bear's Tooth," or the "Tooth," an appellation bestowed by Mr. Fidler on some point which cannot now be identified.*

Other objects which have given names to mountains, on account of real or fancied resemblance of form, and which are of interest because so extensively used, are : the saddle, the sugar-loaf, the needle, the pap or nipple, the hay-stack, and many others. There are several "saddle-backs" in England, one near Skiddaw (2,787 feet) being the most celebrated. Black's Guide says of it : "Blentha-

* See *ante*, p. 49.

cara is the ancient name, which now-a-days is more usually termed Saddleback, an appellation acquired from its shape when viewed from the neighborhood of Penrith." In this country we have reversed this style of nomenclature. "Saddle-back" was long considered a good enough name for the highest mountain in Massachusetts; but of late years, the more elegant one of "Graylock" has been coined for it. The most famous Saddle Mountain, however, is "La Silla," near Caracas, the first high point ascended by Humboldt on the American continent.

There are many mountains which bear the name of "sugar-loaf" in various parts of the world. The conical peak on the west side of the entrance to the harbor of Rio Janeiro — the "Pão de Assucar" — is perhaps the best known of these. There is one near Abergavenny in Wales, which as viewed from the east is "a perfect cone tapering finely to a point at a high angle" (Mackenzie). The best-known sugar-loaf in this country is one of diminutive size, an "eddy-peak" of Triassic sandstone, near Deerfield, in the Connecticut Valley.

The term **needle** as the designation of a mountain is much more commonly used by the French than by the English. It is especially in the vicinity of Chamonix that the clusters of sharply pointed peaks bear the name of "aiguilles." The rocks there are of a peculiar texture (slaty-crystalline), and have an almost vertical position of the cleavage planes. These conditions cause the elevated ridges to weather away under the influence of the erosive agencies, in such a manner as to leave points projecting, either singly or in groups, above the general level, most naturally suggesting by their sharpness the idea of the needle. The "needles" best known in English-speaking countries are the pinnacles of chalk on the extreme western end of the Isle of Wight, a famous landmark for vessels bound to Southampton. They are sharply pointed rocks "which have been produced by the decomposition and wearing away of the chalk in the direction of the joints or fissures by which the strata are traversed" (Mantell).

There are various mountains called the **Paps,** both in England and the United States. The " Paps of Jura " are well-known elevations on the largest of the Hebrides group of islands, visible far at sea and from all the western coast of Argyllshire. The " Paps " on the north shore of Lake Superior are fine rounded summits, surrounded by noble scenery. "Teton," the French equivalent of "pap," is also used with the same topographical meaning as the latter word. The "Teton Range," near Snake River, in Wyoming Territory, is one of the most impressive of all the Cordilleran mountain groups; and the "Grand Teton," 13,691 feet high, is its culminating point. This may have been the mountain called the "Tooth" by Mr. Fidler. It certainly has much the appearance of a gigantic tooth, as seen from one point of view.

There are **Hay-stacks** without number, both in this country and in England. Rising high above Buttermere Water are the "Haystacks" of the Lake District. One point in the Yellowstone National Park is thus designated; and there are various others with the same name in the Appalachians.

Hog-back is not a particularly elegant word, but it is one put to a variety of topographical uses in this country. There is, among others, a Great Hog-back in North Carolina, 4,790 feet high. Along the eastern base of the Rocky Mountains the strata are broken off and upturned in grand crests, producing a most peculiar and picturesque type of scenery, especially attractive to geologists from the clear revelation there afforded of the nature of the mighty forces by which that grand system of mountains has been uplifted. These crests are familiarly known to those living in that region as "hog-backs;" and the belt along the base of the chain where these peculiar forms occur, is called the "Hog-back Country." Furthermore, the remarkable ridges of gravel occurring in Northern New England and elsewhere in this country, known as "kames" or "eskars," and which are such a puzzle to geologists — although by most of them ascribed to the action of ice — are frequently called "hog-backs," as also "horse-backs."

The **Camel's hump** is not as popular topographically as the hog's back; but as occurring not uncommonly, and especially as designating one of the highest of the mountains of Vermont, it should not be omitted from the list of imitative names — a list which might easily be considerably extended beyond that which has here been given.

Among words designating some peculiarity of form in the rocky outcrops which are so often seen in mountainous regions, and especially near the summits of lofty peaks, the following may be mentioned as being in current use wherever English is spoken: "precipice," "cliff," "crag."

Precipice (Lat. *præceps*, headlong, *precipitare*, to fall headlong or head-first, from *præ* and *caput;* Fr. *précipiter, précipice*) is the most general term in English for any very steeply inclined wall or surface of rock. A "defile" is bordered by "precipices;" a "gorge" has "precipitous sides."

Cliff is nearly akin to "precipice;" in fact, there is hardly a perceptible difference in meaning between the two words. "Cliff" is etymologically the same as "cleft," "cleugh," and "clove," coming from the Anglo-Saxon "clif," a shore, a rocky shore, and hence rock, and connected with *clifian*, to cleave or split asunder. A cleft in the rock has precipitous or cliffy sides; hence a cliff is a nearly perpendicular face of rock. Steep faces of sand or gravel are more commonly called "banks" than cliffs. The "bluffs" of the Prairie Region might well be called "cliffs;" and, in fact, a certain limestone which frequently is seen in the Lead Region cropping out and forming bluffs along the streams has been often called, by geologists as well as by the people generally, the "Cliff limestone."

"Cliff" is a favorite word with the poets; and here follow some examples of the use of "precipice," "cliff," and "crag" by Scott, Tennyson, and Lowell: —

> "Seems that primeval earthquake's sway
> Hath rent a strange and shattered way
> Through the rude bosom of the hill,
> And that each naked precipice,
> Sable ravine, and dark abyss,
> Tells of the outrage still."
> *Lord of the Isles.*

> "Adown the black and craggy boss
> Of that high cliff, whose ample verge
> Tradition calls the Hero's targe."
> *Lady of the Lake.*

> "As over rainy mist inclines
> A gleaming crag with belts of pines."
> *The Two Voices.*

> "A heap of bare and splintery crags
> Tumbled about by lightning and frost,
> With rifts and chasms and storm-bleached jags
> That wail and growl for a ship to be lost."
> *Pictures from Appledore.*

Crag is a word of Celtic origin (Gal. *creag*, Welsh *craig*, a rock), much in use in Scotland, and very familiar to us from its occurrence in books, although rarely heard, in this country at least, in actual use, as a designation of any part of the rocky landscape. There are many names of places of which "cliff" forms a part; for example, Undercliff.

Scar is a word quite unknown to us, except through books, as a topographical designation, however familiar it may be with the meaning of a mark left by a wound which has healed over. The word "scar" appears in a variety of forms; for example, "scaur," "scarth," "scaw," "carr." It is related to the Icelandic *skor*, a crack or cut; Gothic *skaer*, a rock, from *skaera*, to cut or shear, Welsh *esgair*, the ridge of a mountain. Wedgwood says of this word : "Originally, a crack or breach ; then especially applied to a cliff, precipice, or broken rock, a fragment." This is a word frequently heard in the North of England, especially in Northwestern Yorkshire, where the limestone cliffs are called "scars." Hence the name "Scar limestone," the equivalent in a topographical sense, but not geologically, of our "Cliff limestone." "Scaw" forms a part of various names in the Lake District, and especially of the Scawfell Pike, which has already been mentioned, and which may come up again under "fell." Any face of rock, cliff, or precipice may be called, in the North of England, a "scar" or "scaw." Although not in such frequent use in Scotland, Scott has it occasionally, as for example, —

"Is it the roar of Teviot's tide,
That chafes against the scaur's red side?"
Lay of the Last Minstrel.

"'She is won! we are gone, over bank, bush, and scaur;
They'll have fleet steeds that follow,' quoth young Lochinvar."
Marmion.

The word **escarpment** is one rather frequently used in this country, with the meaning of "cliff" or "precipitous face of rock." It is closely allied to "scar" or "scaur" in origin, both being referred back to the ancient (Aryan) root "skar," to cut. The definition of "escarpment" given in Skeat—"a smooth, steep decline; a military term"—would not hold good in the United States.*

The loose stones and angular fragments of rock which are so often seen accumulated at the base of the cliffs or precipices from which they have fallen, are known in the Lake District of England and in Scotland as **screes**. This word is not in use in the United States, although it is much to be wished that it were. There are various localities called by this name in the Lake District: for example, the

* "Scarp" (also written *escarp*) and "counterscarp" are the military terms related to "escarpment." A "scarp" is a slope so steep that it cannot be climbed, and "to scarp" is "to cut down a slope so as to render it inaccessible." (Wilhelm's Military Dictionary.)

Screes, Wastwater; Red Screes, Kirkstone; Yewdale Screes, Coniston; Cautley Screes, Howgill. As generally used, the word "screes" is the equivalent of our "talus" or "talus-slope," and "debris" or "debris-pile." **Talus** is the Latin for "heel," and hence a slope, the word not being used in French as the equivalent of "screes," as it is in English, both in this country and in Great Britain. Thus, Geikie speaks of "the long screes or talus-slopes at the foot of every hill and crag," and of "slopes strewn with screes and débris." * "Scree" is used both as a singular and a plural, although generally the latter. **Debris** is also used with the same uncertainty of number, both in French and English, and in the latter with or without the accent.

The debris-piles which stretch along the lower slopes of the ranges in the Cordilleran Region are locally known as **washes**. These accumulations, consisting chiefly of sand and gravel, brought down from the mountains above by currents of water, occur on a grand scale in some places, especially on the east side of the Sierra Nevada and on the west slope of the Inyo Range, opposite Owen's Valley. These "washes" start from high on

* A. Geikie, Scenery of Scotland, 2d ed., pp. 172, 165.

the mountain sides, and spread themselves downward, often with a moderate and quite uniform slope, along the entire length of the ranges, furnishing ample evidence that the precipitation was once much more copious than it is at the present time.

The various forms which are the result of the weathering of the rocks under the influence of atmospheric agencies, and which are sometimes quite remarkable, and interesting not only from the geological but from the scenic point of view, have received appropriate names in the different regions where they are most strikingly displayed.

There are numerous "Towers," "Monuments," "Castles," and "Pinnacles," in the Cordilleran Region, as well as "Tower" and "Castle" Peaks. Sometimes these names are rather fancifully given; but often the resemblance of the rock-mass to the object from which it is named is most striking. Thus the "Pah-Ute Monument" on the summit of the Inyo Range is an isolated columnar mass, extremely regular in form and of grand dimensions (it is eighty-five feet high), so that it is visible from far and wide.

"Tower" in the form of **tor** (Lat. *turris*) is quite a common word in England, although one not at all in use in this country. It means sometimes simply a tower, the work of man's hand, and has no other definition in Latham's Johnson ; but, in fact, it is also frequently used to designate certain curiously shaped masses which have been left as the result of the weathering of various rocks, but especially of the granite in Devon and Cornwall.* Some of these weathered masses are so poised that they can be moved or made to rock on their foundations. They are then called "**rocking stones**," or "**logans**," the latter name being peculiar to England. The "logan" situated in Cornwall, near Castle Treryn, St. Leven, is seventeen feet in length, and has been estimated to weigh about sixty-five tons. " Helmen Tor, on Dartmoor, is a rugged hill composed of blocks of granite, several of which ' rock ' with ease." † Rocking-stones in the United States are of rather rare occurrence. One of marble, near Pittsfield, Mass., is of very large size, being twenty-six feet long, and estimated (by the present writer) to weigh about 200 tons. It is called the "balanced-rock ; " but is no longer to be moved by the hand, although

* See farther on, under " Moor."
† H. B. Woodward, The Geology of England and Wales, London, 1876, p. 411.

it is said, on good authority, that it could be so moved a few years ago.

Certain blocks of sandstone and conglomerate which are strewn over the surface of the ground in Dorsetshire and Wiltshire, England, are familiarly known as **gray-wethers.** They are supposed to be the remains of strata of Tertiary age which once covered the region where these gray-wethers occur. It is from these blocks that Stonehenge and other Druidical circles of a similar kind have been built; hence they have been called also "Druidstones" and "Saracen's stones" or "Sarsens." The name "gray-wether" is supposed to have been derived from the resemblance of the objects in question to a flock of sheep seen in the distance.

III. VALLEYS, GORGES, AND CAÑONS.

Valleys connect the plains with the mountains. We can speak of a plain without necessarily thinking of a mountain as its boundary or limit. But in general, when the word "valley" is used, it is assumed that the region spoken of lies within the mountains or has mountains adjacent to it. There is, however, one way in which we employ the word without reference to any particular topographical form, as when we speak of the valley of a river, meaning its "basin," or the entire area drained by it and its branches. In such a case the usage varies very much according to custom and the size of the stream itself. As ordinarily used, we mean, when we speak of the "valley" of any river, the lower and more level belt adjacent to the stream. One who lived in the North Park near the borders of Colorado and Wyoming, at an elevation of 7,000 or 8,000 feet above the sea, and surrounded by high mountains, could hardly be said to be living in the Mississippi Valley, and yet he would be within the drainage basin of that river, since that branch of the Platte which rises in the North Park is an affluent of the Mississippi.

As the word "valley" is most generally used, it means a depressed area between two mountains or mountain ranges. Every system of mountains is made up, in general, of highlands with intervening low and comparatively level areas in which the drainage from the adjacent slopes is collected, and which widen out as we recede from the higher regions, and finally merge in the plain, or enter the ocean when this is so closely adjacent that there is no room for a plain between it and the sea. "Valley" is the most general term in English for these areas, and it is applied without reference to size or altitude, or even the character of the vegetation.

Valleys are, as a general rule, parallel with the subordinate members of the system of ranges within which they lie. Thus the Great Appalachian Valley runs for hundreds of miles, for long distances varying but little in width or general character, parallel to the Blue Ridge, a very persistent member of the Appalachian

System. Very wide and lofty ranges, like the Alps, have two or more systems of valleys, those of the first order being parallel with the main range, and those of the second order approximately at right angles to this, and occupying depressions between the "spurs" of the central range, these spurs being those elevated portions of the uplifted mass which have most successfully resisted the action of those erosive agencies by which the mountains are being slowly but surely worn away. The word "valley" (Lat. *vallis*, Ital. *vallata*, Fr. *vallée*, Sp. *valle*, etc.) is used almost exactly in the same way in which it is employed in English by the people whose language is directly descended from the Latin. In the Germanic and Scandinavian tongues the equivalent is "Thal" (German) and "dal" (Swedish), which words are closely related to our "dale." **Vale** is simply an abbreviation of "valley." It is a poetical rather than an every-day word; and both "valley" and "vale" are used in various phrases in which the topographical meaning is nearly or quite lost, as in "vale of years" (Shakspeare), and the biblical phrase "valley of the shadow of death," in both of which the gloom of a deep valley seems to be indicated.

As long as we limit ourselves to those simple depressions between mountains to which the words "valley" and "vale" are most commonly applied, we find comparative simplicity in the nomenclature; but as soon as we inquire into the names of those lower areas which cut entirely across the range, or lie so as to be within the elevated mass, or, as it were, to form a portion of it, we find that although these areas do to a certain extent come within the definition of valleys, they are extremely varied in form and character, and that there is a corresponding variety and complexity in the names by which they are designated.

There are several Spanish words, resembling in meaning those which have just been cited, one of which is not only in common use in the Cordilleran Region, but very familiar in all parts of this country, and it has even been carried from America to England, and to the Continent of Europe. This word is **cañon**; others resembling it in meaning are **cañada** and **cajon**; but the two latter are much less familiarly known to us than is the former. "Cañon" is the augmentative of *caña*, a reed, or tube. As used by the Spaniards of Spain, it means a "cannon;" and it is not found in any *Spanish* dictionary with the signification which it has in this country — namely, "a valley, and especially a somewhat narrow valley, with steeply sloping sides; a long, deep ravine or gorge, or even a defile." A river which has been flowing through an open valley suddenly becomes hemmed in between lofty, precipitous, or even perpendicular walls; in the language of the Cordilleran Region it is said "to cañon." But even valleys which are rather broad and open are sometimes called "cañons." *The* "cañons *par excellence* are those of the Colorado Region; and the most stupendous of all is the "Grand Cañon," where the river flows at a depth of 5,000 feet below the general level of the country, and between almost perpendicular walls.

Broad and open valleys are called in Spain "valles;" when they become narrower, and their sides are more precipitous, they are known as "cañadas," and not as "cañons," either in Spain or in any part of South America.* Even in Mexico the use of this word with the topographical meaning so familiar to us is almost entirely unknown. As if to keep us in mind that there is such a word as "ca-

* Barcia defines "cañada" as "el espacio que hay entre dos montañas ó alturas poco distantes entre sí."

ñada," one valley in California bears that name — the "Cañada de las Uvas;" but, in general, throughout California and the Cordilleran Region, all valleys except the very broad ones, and many gorges and defiles are called indiscrimininately "cañons." The word "cajon" means something intermediate between a "cañon" and a "defile" or "pass," and will be noticed farther on.

Very few ranges of mountains preserve a continuity of height for any great distance. Their outlines, as seen from a distance, are "serrated,"* elevated crests, ridges, or peaks alternating with depressions. Some ranges are cut very deeply down by these depressions; in others the difference between the most elevated and the most depressed portions of the range is small. The "crest height" of a chain of mountains is indicated by a line connecting its highest peaks; its "pass height" by one drawn so as to touch its depressions; and ranges very generally maintain something like the same ratio of crest height to pass height for long distances.

* Not "serried," as some writers have it; for example, Hull (in Physical Geography and Geology of Ireland, pp. 142 and 163), who speaks of "serried ridges" and of "rocky and serried aspect." "Serried" means "crowded together," from Fr. *serrer*.

The depressions in a range are called — as we see — **passes**, since by their aid we pass over the mountains. "Peaks, Passes, and Glaciers" is the title of the first series of volumes published by the English Alpine Club. The use of the word "pass," however, generally implies something grand and elevated. The Alps, Himalaya, and the Cordilleras have their "passes;" while this term is comparatively little used in the Appalachians, or in the mountains of England and Scotland. In the White Mountains of New England the passes are occasionally called **notches**, a local use of the word peculiar to this region and that adjacent to it; for example, the "Crawford Notch," the "Dixville Notch." In the Catskills the passes, as well as the valleys themselves, are known as **hollows**. In Pennsylvania and farther south they are called **gaps**. Those which are cut down deep enough to allow the water to pass through from one valley to another are designated as "water gaps;" those which are but shallow notches on the edges of the long straight ridges so characteristic of the Appalachians are called "wind-gaps." The "Delaware water-gap" is a famous locality, where the river of that name breaks through the Kittatinny Range.

Passes in the French Alps are called "cols" (Lat. *collum*, neck), and this word is frequently used by English Alpinists. Thus, Tyndall says: "Crossing the col, we descend along the opposite slope of the chain. . . . If the valleys on both sides of the col were produced by fissures, what prevents the fissure from prolonging itself across the col? . . . The cols are simply depressions," etc., — all within ten lines.* "Neck," in English, as a topographical word, means a narrow isthmus connecting two distinct areas of land, and is a term but little used except near the seashore.

"Passe" is occasionally used in French, and "paso" in Spanish, as the equivalent of "col" and "pass;" but in the French Pyrenees the passes are much more generally termed "ports" (Lat. *porta*, Fr. *porte*, Sp. *puerto*, door), a word which is used in all these languages for "harbor" and "mountain pass," as well as for the ordinary door. In the Andes "paso" means any depression in the crest-line of the chain which permits a passage across it. The diminutive of "puerto" — "portillo" — as generally used, indicates a "pass" through a narrow gorge or

* Hours of Exercise in the Alps, London, 1871, p. 236.

cañon traversing the range. In the Lake District of England a "pass" is also called a door, generally spelled in this case dore (M. E. *dore*, Sw. *dörr*, Ger. *Thor*, door). The passes thus denominated are generally narrow door-like openings between walls of rock, like the Spanish "portillos:" thus, Lowdore, Derwentwater; Mickledore, Scawfell.

The word sty (Dan. *stic*, a ladder, O. E. *sty*, a path, Ger. *Steige*, a ladder, an ascent) is also used in the Lake District for "pass." The pass from Borrowdale to Wastdale is called "Sty Head." Catstycam, often written "Catchedecam," Helvellyn (Wild-cats' ladder hill) is a word the meaning of which will be easily understood, on recalling what has been said about the word "cam" or "kamm."*

Hause, or "haws," is another word used in the Lake District for "pass," and also — like some other words with a similar meaning in other languages — as designating a ridge connecting two higher points, even if not cut sufficiently deep to serve as a "pass." Occasionally a narrow gorge is also called a "hause" or "haws," as, for instance, in the name Haws Bridge, Kendal. This word is closely allied in meaning to the French "col"

* See *ante*, p. 112.

(Lat. *collum*), since it occurs in nearly all the northern European languages in the form of "hals," meaning "neck."

Another word used in the Lake District with the meaning of "pass," or depression in a mountain range, is **swirl** (spelled also "swirrel"), as seen in the names "Swirl Band," Helvellyn, and "Swirl Edge," near Coniston. This word is spelled by the older Scottish poets "swyre" and "swyr," as seen in the following quotations:—

"The soft souch [sigh] of the swyr and soune [sound] of the stremys."
William Dunbar (1450–1520).

"Out owre the swyre swymmis the sops of the mist."
Gawin Douglas (1474–1522).

Professor Veitch explains the last line quoted in these words: "Over the col or neck of the hill, where the summit dips and rises again on the other side, swim high before the vision the wreaths of mist. . . . 'Swyre' or 'sware' is the characteristic word of the Tweed and Yarrow district especially, for the dip [depression] on the summit [ridge, crest] of the hill." * Black's Guide defines the word "swirl" as "a place on the hills

* The Feeling for Nature in Scottish Poetry, Edinburgh, 1887, vol. i. p. 274.

where the wind or snow eddies:" such a place a depression in the ridge would necessarily be.

Still another name for "pass," used to some extent in the Lake District and the Scottish Highlands, is **slack**, which is defined in the glossary to Black's "Guide to the Lakes" as a place "where the tension of the surface is *slackened*, the consequence being a depression; a hollow generally." Professor Veitch, in endeavoring to explain why, among the older Scottish poets, feelings of terror and dread would naturally take precedence of the poetical, and for long dominate over it, says: "Moor, hope, and slack (hill-pass) were associated with deeds of violence, feud, and hostile inroad, and inspired corresponding feelings of dread and repugnance."

In some parts of the French Pyrenees, especially near Mont-Perdu, the deeply cut passes or gorges traversing the range are called "brèches" (M. E. *breche*, A. S. *brece*, fragment, Eng. *breach*, fracture). Thus, Ramond says: "La brèche où conduit le vallon de glace, et qui est en face du Mont-Perdu, est ce que j'appelle la brèche de Touque-rouye." The famous "Brèche de Roland" is that mountain summit in the Pyrenees which is believed to have been split in two by a blow from the sword of the mighty Roland.

Besides the words already given as meaning a mountain pass, there are several others of Spanish origin all of which are more or less in use in Central and South America. **Atravieso** (Lat. *transversus*, Fr. *travers, traverser*, Sp. *atravesar*, to cross, to traverse) is one of these, and is perhaps the most widely distributed of all the terms having this signification. **Boquete** (Sp. *boca*, mouth) is another name for pass heard in the Chilian Andes, and perhaps elsewhere. Plagemann defines it as "a deeply cut gorge, leading directly across a mountain chain." **Collado**, often used as a synonym of "colina," has also sometimes, according to Caballero, a meaning nearly equivalent to that of "pass." *

A "pass" in German is "Joch," yoke, the equivalent of Lat. *jugum*, which is used for the summit or crest of a mountain as well as for yoke. There is somewhat the same confusion in English between "yoke" and "mountain," since we find that in the Lake District the former word is used for "a chain or ridge of hills" (Black's Guide) : thus, "the

---

* He defines "collado" as "sitio que va subiendo en cuesta, por donde facilita su subida y bajada."

Yoke," Troutbeck. The Scottish poet, William Drummond (1585-1649), has as follows:—

> "Fair yokes of ermeline, whose colour pass
> The whitest snows on aged Grampius face."

The summit of the pass is called the **divide**, or **water-shed**. In this last word the "shed" has not the present meaning, but an obsolescent one of "part" or "divide" (Ger. *scheiden*). Skeat says: "The old sense 'to part' is nearly obsolete, except in *water-shed*, the ridge which parts river-systems." The former meaning of this word is illustrated in the following stanza:—

> "O perfite light! whilk sched away
> The darkness from the light,
> And set a ruler oure the day,
> Ane other oure the night."
> *Alex. Hume* (1560-1609).

The "water-shed" of any river basin limits its "area of catchment," as the hydraulic engineers call it. **Portezuelo**, also spelled "portachuelo," is the Spanish for "divide;" and this word is — or was, a few years ago — in current use among English-speaking people in parts of the Californian Coast Ranges.

There are several words in common use in various parts of Great Britain as designating valleys, of which we know little in this country except through English books, and especially English and Scottish poetry, where these names are of very frequent occurrence. Among these words are "dale," "dell," "dean," "dene," and "den," differing little from each other in meaning, and being very nearly the equivalent of "valley" or "vale" as generally used in this country.

Dale (A. S. *dæl*, M. E. *dale*, Ice. *dalr*, Dan. and Sw. *dal*, Ger. *Thal*). Of this word Skeat says: "The original sense was 'cleft' or 'separation;' and the word is closely connected with the verb *deal*, and is a doublet of the substantive *deal*." There seems to be no difference between "dale" and "vale," so far as meaning is concerned, and both are words much affected by the poets. The use of "dale" for "valley" is very common in the North of England. In Northern Yorkshire the valleys are called both "dales" and "gills."

Dell is only a variant of "dale," and like that is a favorite word with the poets. Latham rather prosaically defines it as "a cavity in the earth, wider than a ditch, and narrower than a valley." Here follow some examples of the poetical use of both "dell" and "dale:" —

"High over hills, and low adown the dale."
                    Spenser, Faerie Queene.

            "Anon the shore
    Recedes into a fane-like dell."
                    T. N. Talfourd.

    "Not less the bee would range her cells,
    The furzy prickle fire the dells,
    The foxglove cluster dappled bells."
                    Tennyson, The Two Voices.

    "Would I again were with you, O ye dales
    Of Tyne, and ye most ancient woodlands!"
                    Akenside.

    "A rocky precipice, a waving wood,
    Deep winding dell, and foaming mountain flood,
    Each after each, with coy and sweet delay,
    Broke on his sight."
                    J. K. Paulding.

    "Broad shadows o'er their passage fell;
    Deeper and narrower grew the dell."
                    W. Scott, Rokeby.

"Dell" as used on the Wisconsin River is a corruption of the French word "dalle." Various localities in the Mississippi Valley, and as far west as the Pacific, were long ago called by the French explorers and fur-trappers "dalles," the best known of the places thus designated being the "Dalles of the Columbia," where this river flows over broad sheets of basaltic lava, in a series of cascades. "Dalle" is the French for rock-surfaces of this kind, and nearly the equivalent of our word "flagging-stones," or "flags." The localities in Wisconsin, designated as the "dalles" of the Ste. Croix, Wisconsin, etc., are similar in character to the Dalles of the Columbia, except that the rock is sandstone and not basalt. Along the Wisconsin River, however, near the Dalles proper, are many little side-ravines, curiously worn out by water which has found its way into the numerous fissures or joint-planes by which the sandstone is traversed, and widened them out into a variety of fantastic and picturesque forms. It is to these clefts, ravines, or gorges that the name "dalles" is supposed, by the English-speaking residents of that region, to have been originally applied; and in accordance with this idea the word "dalles" has been changed to "dells."

"Dale" occurs frequently in the North of England as the termination of a proper name; for example, Langdale, Grisdale, Borrowdale, Yewdale, Kendal, etc.

"Dean," "dene," and "den" are names given to valleys in various parts of England, although by no means in as common use as are "dale" and "dell." **Dene** and "dean" (M. E. *dene*, A. S. *denu*, a valley) are not to be found, as topographical words, in the English dictionaries; and **den** is defined in Latham's Johnson only as "a cave, cave of a wild beast;" but both "dean" and "den" are in use in Northwestern Yorkshire as synonyms of "valley" or "dale," and they also occur forming a part of a large number of proper names; for instance, Mickleden, Tenterden, Rottingdean. Wordsworth defines "dean" or "den" as being "in many parts of England a name for a valley."

> "And sweet are the woods and the vales of Dene."
> *W. C. Bennett.*

> "Among thy groves, sweet Taunton Dene."
> *Gerald Griffin.*

The word "den" is of not infrequent occurrence in Scottish poetry; for example —

> "And long and deep shall be my sleep
> In the dowie dens o' Yarrow."
> *Henry S. Riddell.*

> "We'll sing auld Coila's plains an' fells,
> Her moors red-brown wi' heather bells,
> Her banks and braes, her dens an' dells,
>   Where glorious Wallace
> Aft bure the gree, as story tells,
>   Frae Suthron billies."
> *Burns.*

"Dingle" and "dimble" are words not at all in use in this country, but we meet with them occasionally in English books.

They are both variants of "dimple," which latter word is perfectly familiar to us as meaning a "hollow" or "depression," but not a topographical one. **Dingle** is defined by Latham as "a hollow between hills, a dale;" and by Stormonth as "a narrow valley, a glen." These words were formerly more in use than they are at the present time, and their meaning will be made apparent by the following quotations:—

> " Within a gloomie dimble shee doth dwell
> Down in a pit oregrown with brakes and briars."
> *B. Jonson, Sad Shepherd.*

> " In dingles deep and mountains hoar."
> *Drayton, Muse's Elysium.*

> "I know each lane, and every alley green,
> Dingle, or bushy dell, of this wild wood,
> And every bosky bourn from side to side,
> My daily walks and ancient neighbourhood."
> *Milton, Comus.*

> " Yet live there still who can remember well
> How, when a mountain chief his bugle blew,
> Both field and forest, dingle, cliff, and dell,
> And solitary heath the signal knew."
> *Scott, Lady of the Lake.*

Gill (spelled also "ghyll") is a word limited in use to the North of England. It is a word of Scandinavian origin (Ice. *gil*, a deep, narrow glen, with a stream at the bottom; *geil*, a ravine). In Northern Yorkshire "dale" and "gill" seem to be synonymous. In the Lake District a "gill" is a narrow ravine with a rapid stream running through it: as, Dungeon Gill, Langdale; Stock Gill, Ambleside; etc.

Glen is a word of Celtic origin (Cor. *glyn*, Gael. *gleann*), and is defined as meaning "valley," "vale," or "dale." It is a word much in vogue with the poets, and especially with Scott, on almost every page of whose poems it may be found. Here follow some examples of its use: —

"Rough glens and sudden waterfalls."
T. Warton.

"Can silent glens have charms for thee?"
Bishop Percy.

"The summery vapor floats athwart the glen."
Tennyson.

"The buried river rose once more,
And foamed along the gravelly glen."
T. W. Parsons.

In this country the word "glen" is not much used except by the poets. A locality near Greenfield, Mass., which before it had been spoiled by the hand of man was a charming little ravine, through which ran a stream forming various little cascades, was formerly — and perhaps still is — known as "Leyden Glen." "Watkins Glen" at the head of Seneca Lake, N. Y., is another well-known locality, somewhat similar to Leyden Glen, but on a larger scale, and more attractive,

> "With its long chain of headlong cataracts,
> And pools and windings!"
>
> *A. B. Street.*

The "Glen" of the White Mountains is a broad valley, in no respect resembling either Leyden or Watkins Glen, extending along the lower slope of the Carter Range, and having on its western side, in close proximity, the group of five peaks of which Mount Washington is the centre.

A natural transition leads us from such words as "valley," "dale," and "glen," in which — as these terms are chiefly used — softness of outline, beauty, and repose are the characteristic features, to other designations which are associated with scenery having a character of roughness and grandeur. The words "ravine," "gorge," and "defile" are those best known and in most common use among English-speaking people, as names of narrow valleys with precipitous sides.

Defile is derived from the Latin *filum*, a thread or line — as in "file of men," "single file," etc. A "defile" is a narrow passage through which one "threads" his way. This name is most properly given to passes which are of considerable length, and enclosed between high precipitous walls. Such defiles sometimes form the approach to what is properly the "mountain pass," or depression in the crest of the range. There are grand "defiles" in the Afghanistan passes.

Ravine and **gorge** are terms very closely resembling each other in meaning. Both are from the Latin, and both are used by the French very much as we use them. "Ravine" is from the Latin *rapio, ruina* (Ital. *rovina*, Eng. *rapine, ruin, ravine*); and it means originally "rapidity," then "rapidity of a torrent," then "damage or ruin thus caused," and now, more commonly, "the depression scooped out by the ruinous (ravenous, devouring) element." As used at the present time, a " ravine " is something less precipitous and important as a topographical feature than a " defile," and not so grand as a " gorge."

Gorge is the French *gorge*, throat (Lat. *gurges*, an abyss, gulf, or whirlpool, and also stream, or water in general, or even the sea). In English the word is not in common use with the meaning of "throat," although not infrequently employed as a verb, "to be gorged," said especially of animals who have swallowed an inordinate amount of food, and also used in combination, as in the word "disgorge." Shakspeare has : —

"How abhorred in my imagination it is! my gorge rises at it." *Hamlet*, v. i.

"How he hath drunk, he cracks his gorge, his sides."
Winter's Tale, ii. 4.

As most generally used in English, the word "gorge" means a narrow passage, with precipitous, rocky sides, enclosed among the mountains. A "ravine" need not be enclosed by rocks; a "gorge" usually is so enclosed; and this word would hardly be applied to a mere depression in the soil, as "ravine" might be. The poets use the term "gorge" somewhat less freely than they do "dale," "glen," "vale," and "valley." Talfourd has : —

". . . to which a gorge
Sinking within the valley's deepening green
Invites to grassy path."

And Tennyson : —

". . . dark tall pines, that plumed the craggy ledge
High over the blue gorge."

The writer cannot recall a single well-known locality, either in England or in the United States, specially designated as "a gorge," either with or without an addition in the way of a geographical or qualifying epithet. The deep ravine hollowed out by the Niagara River below the falls is sometimes called the "gorge." Lyell, in describing the falls, speaks of it as a "chasm," "ravine," and "gorge," using all three words within the space of nine lines.*

The Spanish equivalents of "ravine," "gorge," and "cañon" — as the latter word is used in this country — are "barranco," "quebrada," and "garganta." There is more or less uncertainty in the use of these Spanish words by various authors, as has been shown to be the case with the corresponding terms adopted by English-speaking people. This is the case, to a certain extent, in Spain itself; but the want of agreement in regard to these, as well as other topographical designations of a similar kind, between authors writing in Spanish in the various Central and South American States, is quite remarkable.

Barranco (written also "barranca") seems to be the most generally adopted word in Spain itself for "ravine" or "gorge." † It is a word of Basque origin (*barruanjo*, to touch bottom, or fall to the bottom). A **barrancal** is a re-

---

* Principles of Geology, 11th ed., London, 1872, vol. i. p. 355.

† Barcia defines this word as "la quiebra profunda que hacen en la tierra las corrientes de agua."

gion of "barrancos," an area deeply furrowed by gorges, a very "broken" country.

Quebrada (Sp. *quebrar*, to break) seems to be the exact equivalent of "barranco." It literally means a "break;" and the adjective form "quebrado" corresponds to our word "broken" as applied topographically, except that it seems to convey the idea of a still rougher country than that simple word would generally be intended to indicate. We should describe a "quebrado" region as one "cut-up" by deep ravines or cañons. "Quiebra" is simply a variant of "quebrada." Raimondi uses the last-named word almost exclusively to designate every possible form of ravine or gorge occurring in the Peruvian Andes.

Garganta (the throat, the gorge) seems, as a topographical designation, to be the exact equivalent of "quebrada" and of the French and English "gorge." Pissis employs it constantly in the topographical description of Chili, intended as an explanation of his great map of that country. While "garganta" occurs on almost every page of that description, the words "barranco" and "quebrada" are rarely, if ever, found in it.

Cárcova is another Spanish name for "gorge," and is apparently nearly the equivalent of "barranco" or "garganta." It is derived from the Latin *concavus*. "Cárcovo" and "carcabucho" are other forms in which this word appears in various works on the geology of Spain; the latter is a form peculiar to a certain district in the Province of Madrid.*

Cajon, as already mentioned, means "defile," "gorge," or "cañon" (as the Americans use this latter word), and especially a defile leading up to a mountain pass; hence also the pass itself. It is the augmentative of *caja*, box. Rivers are sometimes said to be "encajonados," or "boxed in," as we might say in English, when they occupy a narrow valley enclosed between high, precipitous walls; or, as is frequently said in the Cordilleran Region, when they "cañon." The idea of "boxing in" a stream is one not unfamiliar to people living along the eastern base of the Rocky Mountains, since we there occasionally hear of "box cañons"—some narrow defiles with precipitous walls, between which a stream meanders, being thus denominated. There is also a

---

* " Batres con sus carcabuchos (que así llama en quella comarca á los cárcavos ó barrancos de que se halla rodeado)." (De Prado, Descripcion física y geológica de la Provincia de Madrid.)

well-known pass in Southern California called the "Cajon," or "Cajon Pass."

Gully is a word in general use in England and in the United States, with essentially the same meaning in both countries. A "gully" is a very small ravine. A hollow or channel worn in soft earth, gravel, or sand by a heavy rain-fall would be called a "gully;" but a similar one on a larger scale, worn by a permanent stream, would not — as this word is now used — be thus designated. "Gully" appears to be the same word as "gullet" (Lat. *gula*, Fr. *goulet*), meaning throat, neck of a bottle or of any other long-necked article of a similar kind, and hence water-course, which was also formerly its meaning in English. At present "gullet" is used exclusively to designate the throat, and "gully" as a topographical word, with the meaning given above. In the first published attempt at a description of the White Mountains, what we now call "ravines" are spoken of as "gullies." *

The word **gulch** is one in common use in the Cordilleran Region, with almost exactly the same meaning as "gully" in the Eastern States. The smaller ravines worn by water

* Josselyn, New England's Rarities, London, 1672, p. 3.

running down the steep slopes of the river cañons, and dry during most of the year, are familiarly known to the miners as "gulches," and a large proportion of the gold obtained in the early days of California was won by washing the material scraped out of the bottoms of these "gulches" with the aid of the knife or some other equally simple tool. "Gulch" is a good English word, meaning, according to Wedgwood, "a gully or swallow in a river," and closely allied to "gulp," to swallow in a hurry, especially a liquid, and nearly the same as "bolt," a word only used, however, with reference to solid food.

Chasm has come down to us from the Greek, little altered in sound or meaning. It is the Greek χάσμα (from χαίνω, to gape, yawn, or open widely), a yawning fissure, or deep precipitous cavity in the rocks or in the earth in general, and used with this meaning by various classic authors, or just as we use it in English. So rare is its occurrence in French that it is only found in the supplement to Littré, where it is called a "néologisme," and the first use of it in that language credited to Chateaubriand, who in his "Mémoires d'Outretombe" calls the gorge below the Falls of Niagara a "chasme." The word is not found in Grimm's Dictionary in any form. It is occasionally used in England, especially in poetry. Thus Wordsworth, in writing of the "Devil's Bridge," North Wales, says : —

"There I seem to stand
As in life's morn; permitted to behold,
From the dread chasm, woods above woods,
In pomp that fades not; everlasting snows;
And skies that ne'er relinquish their repose."

The word "chasm" has obtained a firm hold as the name of the deep gorge of the Au Sable River, near Keeseville, N. Y., which is now almost exclusively known as the "Au Sable Chasm." Some of the more prominent fissures or ravines worn by the waves in the rocky cliffs of the New England coast are called "chasms;" others are designated as "purgatories" (see that word, farther on). "Rafe's Chasm" near Gloucester, Mass., is a much visited locality, not differing essentially from the Newport "Purgatory."

Gulf is a word which is variously applied in topography, and comes from the Greek κόλπος, κόλφος, the exact equivalent of the Latin *sinus*, meaning the bosom, or a bosom-like fold in a garment, any bosom-like hollow or indentation in the sea-coast, and also (but more rarely) a depression or hollow in the land, a valley or vale. The Latin word *sinus* is also used with all these meanings, excepting perhaps the last. Κόλπος has taken two forms in French — "golfe," and "gouffre." The former is almost exclusively used as in English to designate a deep indentation in the sea-coast. "Gouffre" is more like our "abyss" (Gr. ἄβυσσος, bottomless, unfathomable), and, like that, a favorite with poets, orators, and others who delight in the use of resounding words of rather vague meaning. Any locality in regard to which little or nothing is known, but which is believed to be a place of horror, filled with fire or water, deep, dark, and awful, may be called in French a "gouffre," or in English an "abyss," or "abysm," as Shakspeare has it; thus, from various French authors, "gouffres éternels," "gouffre infini du néant," "gouffres du trépas," etc. — localities the precise situation or topographical character of which it would be hard to define. Quite similar to this is the use, in English, of the words "gulf" and "abyss;" thus, "gulf of torments," "abysm of time," "abysm of hell," etc. As a topographical designation, on land, the word "gulf" is occasionally used in this country. The long, narrow, but deep excavations worn by the streams in Northern

158 TOPOGRAPHICAL NOMENCLATURE.

New York, west of the Adirondacks, are locally known as "gulfs;" thus, the "Gulf of Lorain," the "Gulf of Rodman," etc. In the White Mountains this word is also in use, to a limited extent, as the equivalent of a deep, precipitous ravine; for example, Oakes's Gulf, on the east side of Mount Monroe. The "purgatory" at Great Barrington, Mass., is now sometimes called the "ice-gulf." A similar topographical use of this word, as designating a feature of the land, is seen in the following quotation from James Beattie:

> " Fancy a thousand wondrous forms descries.
> More wildly great than ever pencil drew —
> Rocks, torrents, gulfs, and shapes of giant size,
> And glittering cliffs on cliffs, and fiery ramparts rise."

Flume, a word which was once in use in England as the equivalent of "river" (Lat. *flumen, fluere*, to flow), dropped out of the language, as spoken in that country, and is not to be found in any edition of Johnson. It is, however, a well-known word in the United States, where it means, as ordinarily used, an artificial channel of boards in which water is carried for any purpose, but especially for turning a water-wheel; in English, a "mill-race." A "flume" in California, and on the Pacific Coast generally, is a structure of boards by the aid of which the water of any stream is diverted from its channel for the purpose of washing the sand and gravel in the bed thus left dry.* The wooden aqueducts in which the water of the "ditches" (small canals built to convey water from the mountains for mining purposes) is conveyed across valleys or ravines, are called "flumes." They were found very useful in the "early days" of California for hanging up criminals convicted in Judge Lynch's court. Hence the phrase "gone up the flume," a euphuism extended so as to designate one who has come to grief in any way — hanging included. "Flume" with a topographical meaning is also a familiar word in this country. It is chiefly used in the White Mountain region, where it is the name of several deep, narrow ravines or gorges, with nearly perpendicular walls, through which runs a stream of water forming a series of cascades. By far the best-known locality bearing this name is that situated on a branch of the Pemigewasset River, in the Franconia Notch, in which a huge bowlder formerly hung suspended high above the stream; but which was swept away by the rush of water through the gorge resulting from

* This operation is called "fluming" the river.

a "cloud-burst" which took place on the mountains above, June 20, 1883. The "New Flume," near Franconia, is another locality of very similar character.

A word which has been curiously diverted from its original theological meaning into a topographical one is **purgatory**. Along the coast of New England, and in the interior, narrow ravines with nearly perpendicular walls are called "purgatories." The origin of the name is easily understood. Purgatory (Lat. *purgare, purgatio*) is supposed to be a place not easily gotten through, and not traversed with comfort by those who do succeed in doing this. The topographical "purgatories" are more or less blocked up by huge angular rocks which have fallen from above, and hence a passage through them is no easy matter. The best-known localities bearing the name of "purgatories" are those at Sutton and Great Barrington, Mass., and there is one on the sea-shore at Newport, R. I. The Sutton Purgatory is three or four hundred feet long, about fifty feet wide, and from fifty to seventy high, with nearly vertical walls of gneiss, the bottom being without running water, and covered with large, angular masses of rock. The Great Barrington Purgatory is almost a fac-simile of that at Sutton, except that there are more trees and undergrowth in the latter than in the former. There is also a "purgatory" in the Rocky Mountains, this name being given to a gorge, defile, or cañon, traversed by one of the branches of the Arkansas (Purgatory River). This "purgatory" is on a grand scale, it being more than fifty miles long, and its walls from eight hundred to a thousand feet high.* There are also at least

* Colonel Dodge thinks that this river was called Purgatory River because it had been previously named "Rio de las Animas Perdidas" by the Spaniards, "purgatory" being, as he thinks, the translation of "animas perdidas" (lost souls). This name is said to have been given in consequence of the loss (in what manner is not stated) of an entire Spanish regiment in the cañon. Evidence of the truth of this supposed calamity the present writer has not been able to find. When we consider how completely this locality agrees (topographically) with those which have been thus designated in various other parts of the country — except that it is on a grander scale than the others — the explanation of the origin of the name given by Colonel Dodge seems hardly admissible. Emory and Abert, U. S. officers, the earliest scientific explorers of this region, call this river and cañon "Purgatory River" and "Rio Purgatorio." Colonel Emory adds that the river is also called "the Picatoire, a corruption of Purgatoire." Mr. W. A. Bell, who also explored this region in connection with one of the Pacific Railroad surveys, says, after describing the wonderful effects of color in the cañon of the Purgatory: "There cannot be a doubt that, coming unexpectedly upon this marvellous spectacle, *Purgatory* was the constant and

two localities on the Scottish coast which are designated as "purgatories"—one on the Orkneys, and one on the Shetland Islands.* Of the topographical character of these localities nothing is known to the present writer.

Close to the Orkney purgatory is a place called "Hell;" and this naturally leads to the mention of the fact that "hell" and "devil" play quite a conspicuous part in the topographical nomenclature of this country. There is on the State Geological map of Kentucky a locality called "Hell for Certain," which closely corresponds with a Californian name "Hell Itself," near which latter place is one designated as "Nearly Hell." The two latter names are not on the map, and the present writer supposes them to have been given for moral reasons; the Kentucky name was probably connected with the undesirable physical character of the locality, as is certainly the case with various other places called "hells"

unvarying idea impressed upon the imagination of the French explorers from Louisiana who first visited this spot; for it seemed only just out of some mighty furnace," etc. (New Tracks in North America, London, 1870, p. 88.)

* J. R. Tudor, The Orkneys and Shetland: their Past and Present State, London, 1883, pp. 361, 473.

in the Cordilleran Region. The "devil" is also largely mixed up with topography all over the United States, as well as in Europe. There is hardly any kind of topographical feature which has not somewhere the devil's name attached to it. But his Satanic Majesty is especially well provided with "pulpits" and "slides:" to the former his name is given apparently on theological principles; to the latter as a practical illustration of the familiar phrase " Facilis descensus Averni."

There are several words used to designate a certain peculiar topographical feature seen in various mountain regions. The most easily defined of these is **cove** (Lat. *cavus*, hollow). This word is applied topographically both to the sea-coast and to the mountains. A "cove," as a marine term, is a recess, small bay, or hole in the coast-line, as the "Cove of Cork." Small indentations in the coasts of lakes are frequently designated as "coves:" there are many such, especially on our Great Lakes. A "cove" is also a recess, hollow, or nook among the mountains. This use of the word is common among the foot-hills of the Blue Ridge, in Virginia: it is not at all infrequent in the Lake District of England; for example, Red Cove, Keppel Cove, etc.

"Cove" is occasionally used by the poets, as in Wordsworth's "Excursion"—

"The coves, and mountain steeps and summits."

Another word of Celtic origin, familiar in England, but rarely, if at all, employed in this country, is **coom**, a term spelled in a variety of ways. The original Welsh is "cwm" (pronounced "coom"), and it is thus frequently written by the English. Other spellings are "coomb," "combe," and "comb." * A "coom" is thus defined by Mackintosh, in writing of the escarpments of the North and South Downs: "In most places they [the escarpments] are indented by bays and coves. The latter, in many, if not in most instances, are not valleys, but curvilinear recesses, bounded all round by steep slopes — the innermost part of the slope being often the steepest. The coves are sometimes so geometrically curvilinear as to suggest the idea of having been literally whirled out by the eddy of a powerful current. . . . The coves or *cwms* (as I shall hence-

* "Coom" is preferred by the present writer as the spelling of this word, because it seems best to spell as pronounced when there is no special reason to the contrary. "Cwm" is an unpronounceable combination of letters to those not acquainted with Welsh.

forth call them) are not confined to the escarpments." * These cooms form one of the most characteristic features of the scenery of a great part of North Wales, and are more or less common in other parts of Great Britain. In Northern England " coom " has very much the same meaning that it has farther south. Black's " Picturesque Guide to the English Lakes " defines this word as " a hollow in the side of a hill." As an example, Gillercoom, Borrowdale, may be mentioned. As the terms are used in the Lake District, there seems to be little, if any, difference between a " coom " and a " cove."

Corry is another Celtic word, equivalent in meaning to " coom." It is the Gaelic *coire*, meaning a " caldron " or " large kettle," a name applied, as Ramsay says, " to those great cliffy semicircular hollows or *cirques* in the mountains in which tarns so often lie." † Kinahan says: " In connection with the hills are the cooms or corries, which are more or less rounded, bowl-shaped hollows or valleys enclosed, on all sides but one, by steep and

---

* Mackintosh, Scenery of England and Wales, London, 1869, p. 98.

† Physical Geography and Geology of Great Britain, 3d ed., London, 1872, p. 285.

in some cases perpendicular cliffs. In SW. Kerry the cooms are very numerous, and of great dimensions, some of their bounding cliffs being over 1,000 and 2,000 feet high."* In Wales there is a lake on Cader Idris called Llyn Cyri (pronounced *curry*), a name unintelligible to the Welsh, but a remnant of the Gaelic (Ramsay). Sir Walter Scott has it:—

> "Fleet foot on the correi,
> Sage counsel in cumber,
> Red hand in the foray,
> How sound is thy slumber!" †
> *Lady of the Lake.*

Wordsworth thus describes a "corry" in the Cumbrian Chain:—

> "A little lowly vale,
> A lowly vale, and yet uplifted high,
> Among the mountains;
> Urn-like it was in shape, deep as an urn,
> With rocks encompassed, save that to the south
> Was one small opening, where a heath-clad ridge
> Supplied a boundary less abrupt and close."
> *The Excursion.*

"The word "coom," in the form of "combe" and "come," is widely spread over France and Switzerland. In Burgundy and some parts of the Morvan, "combe"

* Kinahan, Manual of the Geology of Ireland, London, 1878, p. 309.

† In the note to the above it is added: "or *corri*, the hollow side of the hill, where game usually lies."

means a "valley," "gorge," "depression," or "cove" in the mountains; hence, "level land," especially such as is used for pasturage. We find in the environs of Autun the name "Comberland" as the designation of a small estate of pasture-land, the origin being evidently analogous to that of the English "Cumberland." In the Nivernais and in Burgundy a large number of places bear the name of "come" in various forms; for example, Comeau, Comaille, Comagne.*

In the Jura the longitudinal valleys are called "combes." They are depressions which have been formed, sometimes by actual longitudinal disruption of the rocks, but more generally by denudation, which has acted unequally on rocks of different geological character. Many of the great valleys of the Alps are of this type; they occur usually along the line of junction of a hard, crystalline rock with one which is soft and stratified.

The early appearance of the word "combe" in Latin (as early as the seventh century) and its wide-spread distribution over Europe have led to considerable discussion as to its etymological relations. Du Cange thinks that it is

* Chambure, Glossaire du Morvan.

from *cymba* (boat), in allusion to the boat-shape of some valleys; Diez inclines to derive it from Lat. *concava;* Littré prefers a Celtic origin for the word. Skeat says: "The original sense was probably 'hollow;' *cf.* Gr. κύαρ, a cavity." The Aryan root suggested is *ku*, to contain, to be hollow.

While the longitudinal depressions in the Jura are known as "combes," the deep, transverse gorges are called **cluses**. These often cut the ranges vertically, and are considered to have resulted from the occurrence of great fissures or "faults" (as geologists call them) by which the rocks have been actually "rent in twain." Hence the word "cluse" has been rendered into English as "valley of disruption."* "Cluse," as a topographical designation, is not in use in England, although we have many words derived from the Latin *claudo*, *cludo*, from which come *clausum* and *clusum*, an enclosed space. The Germans have this word in the form of "Clause," defined in Grimm as "fauces montium," jaws of the mountains, and most commonly used to designate a mountain pass capable of being defended against military attack.

* See Ball's Alpine Guide.

"Cleugh," "clough," and "clove" are all variants of the same word, which is also spelled in various other ways,—for example, "clew," and "cleuch," in some older works. It is related to the very frequently used word "cliff." Skeat says of "clough": "An English form with a final guttural, corresponding to Icel. *klofi*, a rift in a hill-side, derived from Icel. *kliúfa*, to cleave. Similarly *clough* is connected with A. S. *cleófan*, to cleave; and is a doublet of Cleft." The word "cleugh" is frequently heard in Southwestern Yorkshire, where the rugged glens are called "cleughs." Professor Phillips says: "These branches [of the Calder] frequently descend through rude and craggy fissures, to which the name of 'clough,' replacing 'dale,' is applied." * Geikie defines a "cleugh" as "a still narrower [than a "dale" or "glen"] and steeper-sided valley, chiefly to be found in the higher parts of the uplands." He further adds: "A 'hope' is the upper end of such a narrow valley, encircled with smooth green slopes." † This word **hope**, nearly the equivalent of "coom" and "corry,"

* The Rivers, Mountains, and Sea-Coast of Yorkshire, London, 1855, p. 96.
† The Scenery of Scotland, 2d ed., London, 1887, p. 303.

seems limited in use to the Southern (Scottish) Uplands. **Clove** is the Dutch word "kloof" (cleft), and is frequently heard in the Catskills, where the deep and wild gorges are called by that name, one of several relics of the topographical nomenclature of the early Dutch settlers of that region.

The "combes" of the Jura sometimes terminate in amphitheatral forms, like the Welsh "cooms," but on a still grander scale. One of the finest of these is the famous "Creux du Vent," or Hollow of the Winds, "creux" (cavity) being one of the names given by French-speaking people to these peculiar topographical forms.

The Pyrenees exhibit this remarkable scenic feature on a grand scale, the so-called "Cirque de Gavarnie" being probably the most striking of these "cirques," or "amphithéâtres" as they are also called in French. This is an "amphitheatre" of which the steps are of gigantic size, snow-covered, and overshadowed by stupendous mountain summits. Another local name for an amphitheatre of this kind is "oule," from the Spanish "olla," a pot or kettle.

For the same reason that these "cirques" are sometimes called "oules," the Geological

Survey of California, while exploring the Southern High Sierra, gave the name of "Kettle" to one of these grand amphitheatral depressions, which so strikingly resembled a kettle in form that it was impossible to refrain from applying that name to it. A gigantic kettle it is, however, for its edges are from 1,100 to 1,600 feet above its bottom, which latter is smooth and rounded in the most perfect kettle fashion.

The so-called "creux" of the Channel Islands are quite different in character and origin from those of the Jura Mountains. They are related to caverns, blow-holes, and purgatories. For instance, the so-called Creux Mahie, the largest cavern in Guernsey, is entered from the sea-shore, near Corbière, by a narrow opening, nearly closed by blocks of rock; but when once the visitor is fairly inside he finds himself in a cave 200 feet long, from forty to fifty feet wide and about the same in height. These "creux" are due to the combined action of the ocean and of atmospheric agencies. The soft and easily decomposed material, forming veins by which the rock is traversed, is worn away above, at the general level of the region, by the action of the rain, and lower down by that of the waves. Hence there results a great variety of more or less opened fissures, clefts, and caverns. When the fissure is entirely open from the general level of the surface to the edge of the sea, we have a "purgatory"—although not so called in the Channel Islands; when there is communication through from top to bottom, but covered for a portion of the distance, the result is a "blow-hole;" if there is an extensive widening of the fissure, as is the case in the Creux Mahie, a cave is the result; but these various forms are all locally designated as "creux."

"Blow-hole" is the name given to these partly covered fissures on the west coast of Ireland and Scotland, as well as on the north side of Cornwall. Sometimes, in heavy gales of wind, the sea is driven into these holes and forced out at the top in grand masses of foam and spray. Hence these "blow-holes" are also called "puffing-holes," and "bullers" or "boilers."

IV. PLAINS, PRAIRIES, AND SAVANNAS.

THE nomenclature of mountains and parts of mountains depends — as will have been seen from what has been already stated — chiefly on *form*. As soon as we begin to consider the level, or approximately level, portions of the earth, and to study the various names by which these are known, we perceive that the *character of the vegetation* plays an important part in the nomenclature. A few words in regard to this may therefore appropriately precede the more detailed enumeration and explanation of the names belonging to this division of our general subject.

The surface of the earth is very unequally covered with vegetation, and this vegetation is of a very different character in different climates, latitudes, and elevations. In general, warmth and moisture are favorable to vegetable growth, and for this reason the tropical regions are those where we expect to find the greatest luxuriance of plant life. The extreme northern and southern land areas are, on the other hand, almost or quite destitute of vegetation, because, although moisture may abound, the mean temperature is too low. For the same reason we find that as we ascend high mountains the forests disappear; then the grasses and herbaceous plants; and finally, if the elevation be sufficient, we come to rock, either bare or sparsely covered with the lowest forms of vegetable life, while still higher these give place to eternal snow and ice. The portions of the earth's surface which are nearly or quite destitute of vegetation form but a very small part of the entire land area, and are of little importance from the present point of view. It is the presence or absence of forests, and their peculiar distribution, which is the most important element in the nomenclature of the level portions of the earth; but where trees are wanting, then the character of the shrubby or grassy vegetation may be more or less clearly indicated in the name applied to the region in question. Thus nomenclature becomes, when we have to do with the more level portions of the earth's surface, largely a matter of botanical geography and climatic peculiarities, rather than of form.

An examination of various regions, either by personal inspection or by the study of botanico-geographical maps, shows us that a very considerable portion of the land is destitute of forests, and that this is often the case where there is no lack of warmth, and also — although to a much more limited extent — where the conditions of both temperature and moisture appear to be favorable to the growth of an arboreal vegetation. Further examination shows us that these non-forested regions are, in very large part, the more level areas — the plains, prairies, steppes, llanos, pampas, campos; in every one of these words the idea of the absence or scarcity of trees is connected with that of a level or slightly undulating surface.

Without going into minute detail as to why this is so, a few hints may be given throwing light on the question, and which will be of service in a further examination of the meaning of the words which have just been mentioned, as well as of others of somewhat similar character.

Absence of sufficient moisture is by far the most important agent in checking the development of forests. A comparison of maps showing the position of the isohyetal curves throughout the world, with those indicating the character and distribution of the arboreal vegetation, furnishes the most convincing evidence of this. In the United States, for instance, we find the region of the "Plains" to be that where the annual precipitation falls below twenty or twenty-five inches. Hence the interiors of the great land masses are most likely to be treeless regions, because the borders of the continents, as a general rule, receive more rain than their interiors. When these borders are mountainous, and especially when the mountains lie athwart the direction of the prevailing winds, they cut off the precipitation almost entirely, so that, in going but a very short distance, we may pass from a region of excessive rain-fall to one of extreme aridity.

But there are other causes besides cold and dryness which are unfavorable to the development of forests. The physical character of the soil is one of the most important of these causes; and the present writer has elsewhere shown the truth of this statement, and furnished abundant evidence that extreme fineness of the soil is the chief cause why extensive regions, otherwise favorably situated for the growth of trees, are destitute of them.* With these facts in view, it is easy to see why plains are more likely than mountain slopes to be treeless. It is toward the plains that the finer material abraded from the higher regions is constantly being carried. The farther from the mountains — that is, the broader and more extensive the plain — the finer will be the material deposited upon it; and this is true whether the detritus thus conveyed be laid down as a subaerial or submarine deposit. In a mountain and plateau region, like much of that between the Rocky Mountains and the Sierra Nevada, we find the more level portions almost entirely destitute of forests, while the mountain ranges which extend across that part of the country are to a certain extent timbered, partly because they are high enough to condense some of that moisture which does not fall on the lower regions, and partly because the finer material, inimical to the growth of trees, has been swept down the steep slopes into and over the broad valleys which lie at the bases of the mountain ranges.

* See Geology of Iowa (1858), vol. i. p. 23; The American Naturalist for October and November, 1876; Science for All, vol. v. p. 124; Encyclopædia Britannica, 9th ed., vol. xxiii., art. "United States."

These preliminary remarks will enable the reader to understand why, in the nomenclature we have here to investigate, we find so generally that the topographical designations include with the idea of flatness and absence of mountains that of a corresponding absence of forest vegetation. We shall also see how it is that a number of foreign words have become very familiar to us as appellations of regions in other parts of the world exhibiting certain peculiarities of surface and of vegetation, and why these same words are so frequently used by travellers and writers on physical geography in their descriptions of our soil, climate, and scenic features.

The most general and commonly used word in English for a level area is plain (Lat. *planus;* It. *piano, pianura;* Fr. *plaine*). A "plain" may be either large or small, forested or bare of trees, or covered with a shrubby vegetation; it may also be low or elevated. It is the antithesis of "mountain." The land surface of the earth consists of mountains and plains. Besides this use of the word "plain" in a general way as the opposite of "mountain," we find that there are interesting specializations of it both in England and in the United States. The most important is in this country; the vast, nearly level area extending west from a little way beyond the Mississippi to the base of the Rocky Mountains being now generally known as the "Plains." "The Plains of the Great West" is the title of Colonel Dodge's work giving his experience in that region. "Life on the Plains" is to us a familiar phrase. No American would have any difficulty in understanding what is meant by "the Plains." The term has come gradually into use since the days of Lewis and Clarke, who at the beginning of their journey hesitated whether to call the treeless portions of the region over which they were travelling "prairies" or "plains," but who soon dropped the former term. Since their day the "Prairies" and the "Plains" have been distinctly separated from each other, and among travellers in the West there is no confusion of the two terms. The plains of England are on a small scale as compared with those in this country. The broad, undulating, treeless areas underlain by the chalk, and forming a sort of table-land to the west and north of the London Basin, are generally called "downs" (see farther on); but there is one well-known locality designated as a "plain" — Salisbury Plain — a name familiar to many as the home of the "Shepherd of Salisbury

Plain," a pious tract once (and perhaps still) extensively circulated in this country.

Plateau and **table-land** are nearly synonymous terms — the one French, but now thoroughly Anglicized, the other English. These words carry with them the idea of elevation and extent. They are scientific geographical designations rather than such as are used in every-day life. The elevated comparatively level regions on which great chains of mountains, like the Himalaya and Cordilleras, are built up, are called "plateaux." By "Plateau Region," in this country, we mean the vast area extending from the Rocky Mountains west to the Sierra Nevada and Cascade Range — a region having an elevation of from 3,000 to 6,000 feet above the sea-level, and on which are built up numerous ranges of mountains, some of which lack little or nothing of being as grand as the Pyrenees. The great uplifted flat areas of land, separated by the cañons of the Colorado and its branches, are called "plateaux," as the type of which the Kaibab may be taken, 7,500 to 9,300 feet high, quite flat on the top, and isolated almost entirely by gorges thousands of feet deep. The word "plateau" is of rather recent introduction into the English language; it is not found in the earlier editions of Johnson's Dictionary, down to and including that of Todd (1827). In Latham's edition (1876) it is defined simply as "table-land." Central Asia is peculiarly a region of plateaux. The stupendous mountains of that portion of the Continent rise from equally stupendous plateaux. The Pyrenees, Alps, and the Caucasus, on the other hand, are mountain regions almost wanting in these broad elevated plains or plateaux.

The flat summits of mountains are sometimes called "tables," and especially in California, where there are several "Table Mountains," all fragments of great lava-flows, capped usually with horizontal or table-like masses of basalt. The "Table Mountain" of South Africa is, however, the best known of the eminences thus designated, and is the only one furnished with a *table-cloth*.* There are two tabular hills forming conspicuous landmarks on the northwest side of Skye (one of the Outer Hebrides) which are known as "Macleod's Tables." Like the Californian "Table Mountains," they are capped with horizontal beds of lava.

* The cloud of vapor borne in from the sea and condensed on the summit of Table Mountain.

Highland and table-land are by no means synonymous terms. Flat regions are, as a general rule, not called "highlands." Certain mountainous districts have almost a monopoly of that name, as the "Scottish Highlands," and the "Highlands of New York," which latter is the designation of the precipitous ranges through which the Hudson River finds its way in the vicinity of West Point, and which is thought to be one of the most picturesque spots in the Atlantic States. The whole Cordilleran Region has sometimes been called the "Western Highlands;" but this name has not been received with favor.

The Spanish use the word **mesa** (table), and its diminutive **meseta**, not exactly as we do its English equivalent, but rather to designate broad **terraces**, as we call them — a river being said to be "terraced" when we rise from it on either side, not by a gradually ascending slope, but by a succession of steps or steep inclines, between which are comparatively level areas. This is a topographical condition of very common occurrence throughout the world, and it is one of the many existing evidences of the much greater volume of the rivers in former times. Each steep rise, with its corresponding level area above, is called a "terrace," or, in Spanish, a "mesa." In the Colorado Region these terraces occur on a grand scale; and that part of the Southwest is frequently called the Mesa Country, or Region, many Spanish names being still current there. While "mesa," as often used, is nearly the equivalent of "terrace," "meseta" has more frequently the meaning of "plain" or "table." A flat area of moderate dimensions, occurring in a mountainous region, and not forming a part of a river bottom, would, by many writers, be called a "meseta."

While the use of the word "plain" does not necessarily imply that the region thus designated is destitute of trees, yet it would generally be understood that this was the case, unless the contrary were especially stated. There are words, however, which while conveying the idea of flatness of surface, also distinctly include that of entire absence or decided scarcity of forests. The word of most importance in this connection, because most widely and extensively used by writers in English, is **savanna**, spelled frequently, especially in older books, "savannah." This word has come to us from the Spanish, and yet with its present signification it is decidedly

American. It is the Spanish *sabana* (sheet), which originally had the accent on the first syllable, and is believed to be derived from the Greek σάβανον, which again is thought to be connected with the Arabic *sabaniya*, from Saban, a place near Bagdad, where linen is, or was formerly, made. The Latin form of this word was originally *sabanum*, and later *sabana*, in which form it appears as early as 781 (Littré). Its first use in Spain as a topographical designation seems to have been exclusively with reference to snow or ice, just as we say in English "a *sheet* of ice." It appears with this definition in the first edition of the Dictionary of the Spanish Academy, (1739), and with the accent on the first syllable.* It then disappears from subsequent editions, down to as late, at least, as that of 1822, but is found in all the later Spanish dictionaries, including those of Caballero (1865), Salva (1865), Dominguez (1882), Barcia (1882), with the accent always on the second syllable, and is defined as meaning "an extensive, treeless plain (*llanura*)," and generally with the additional remark, either that this is a word much used in America, or

* "Por semejanza se llama el plano grande nevado que esta mui blanco e igual."

that the plains called "sabanas" lie west of the Mississippi.* With this topographical meaning the word in question seems to have been first used by Oviedo, who in his "Historia de las Indias" (1535) speaks of a certain region in the West Indies as being a "tierra de muy grandes savánas é arroyos muchos."

This word also appears at an early date in French (*savane*), in works describing the geography of portions of the American continent where this language was spoken. It is defined as being the equivalent of "prairie" (meadow) in Pelleprat's dictionary (1655). Curiously enough, the same word (*savane*) appears also early in the history of this country as meaning, in Canada, not a dry, treeless plain, but a low swampy region covered with a tangled and dwarfed forest growth. This is

---

* Barcia (1882) says: "Sabana — Páramo, llanura sin arboles, extensa y arenosa. Es voz de mucho uso en America." In quite recent official works on the geology of Spain, written and published in that country, the word "sabana" is occasionally used just as it has so long been in America. Thus the great central, treeless plain of Spain is called the "gran sabana central" (not sábana), and other treeless areas are designated as "sabanas." The same word is also, although rarely, applied to the water, it being used exactly as we do the word "sheet" when we speak of a "sheet of water."

the definition of *savane* given by Charlevoix;*
and the same still holds good in that part of
the country, for the writer, while surveying
the Lake Superior region, always heard his
French *voyageurs*† call what we designated
as " cedar-swamps " — horrible, tangled,
swampy thickets of arbor-vitæ — by the name
savane. Charlevoix adds that the dwellers in
the *savanes* are known as *savanois*.

The word "savanna" (or "savannah") occurs in English books of an early date. Wafer
(1699) uses it repeatedly in describing the
Isthmus of Panama, and in such a way as to
leave no doubt that he meant to indicate by
it a region destitute of forests.‡ Shelvocke
(1726) has no other term for a treeless region
than "savanna." In later years this word
has become less prominent, although still not
unfrequently used by geographers especially
in describing American localities. Humboldt
does not use it in his "Ansichten der Natur,"
in his famous chapter "Ueber die Steppen und
Wüsten," although it occurs in the notes

* Histoire de la Nouvelle France (1744), vol. iii. p. 181.
† Boatmen and packers, old servants of the Hudson's
Bay Company.
‡ "An open savannah." "Savannah on the Westside;
though the Eastside is woodland." (Wafer, Voyages,
etc., pp. 190, 72.)

thereto, with the definition "Grasfluren" (grassy plains or meadows) appended. Humboldt was not acquainted with the distinction made in this country between "plains" and "prairies," and it is not surprising that the English translator (Mrs. Sabine) should have rendered "Grasfluren" by the word "prairies," when the author himself would have put "plains" had he been as familiar with the physical geography of North America as he was with that of the southern division of this continent. At the present time the word "savanna" is — so far as the writer has noticed, in many years of travel — never used in familiar conversation as designating any portion of the treeless area of the Western or Cordilleran Region. It seems, however, to be to a certain extent current in the extreme South, especially in Florida, where the rich alluvial flats along the streams are known as "savannas." Thus Mr. Barbour says, in speaking of the land adjacent to the St. John's River, above Lake Monroe, "it is a flat, level region of savannas, much resembling the vast prairies of Illinois. . . . These savannas are everywhere covered with luxuriant growths of marshy grasses and maiden cane, with occasional clumps of timber, consisting sometimes of but three or four trees, and sometimes being several acres in extent." *

An interesting and important word, especially throughout the Mississippi Valley, is **prairie** (Lat. *pratum*, L. Lat. *prataria*, Ital. *prateria*, Fr. *prairie*, and formerly *praerie* and *prérie*, meadow, pasture-land). This French word is one perfectly familiar to us, and yet not to be found in English dictionaries published in England.† It came into use in the Mississippi Valley through the French missionaries and the *employés* of the Hudson's Bay Company. Father Hennepin (1697) describes the prairies of Illinois so minutely and correctly that a better description could hardly be made at the present time. Lewis and Clarke use the word frequently in the earlier portions of their adventurous journey; farther on, however, they are more inclined to speak of the treeless regions through which they passed as "plains." The distinction between "prairie" and "plain" is one which has come gradually into existence as the routes of explorers and settlers have extended themselves farther and farther west. Every one

* Florida for Tourists, Invalids, and Settlers, p. 31.
† It is not in Todd's Johnson (1827), nor in Latham's edition of the same (1876).

knows that the "Prairie States" are those lying contiguous to the Mississippi, and every one understands what "the Plains" are, and no Western man would ever think of confounding the two designations. Those who have studied the prairies and the plains know very well that the causes which have brought about the treeless condition of portions of the region of large precipitation in the States bordering on the Mississippi cannot be the same as those to which is due the scarcity of forests in the arid belt lying still farther west.

Besides the words "prairie" and "plain" there are two others in use in the Cordilleran Region, with meanings quite limited to certain districts. Very early in the history of the exploration of the Rocky Mountains certain comparatively level, grassy areas were designated as **holes**. Pierre's Hole, "a valley about thirty miles in length and fifteen in width, bounded to the east and south by low and broken ridges, and overlooked to the east by three lofty mountains, called the Three Tetons, which domineer as landmarks over a vast extent of country," * was a noted *rendez-*

* W. Irving, The Rocky Mountains, vol. i. p. 76. The name given to the "eventful valley of Pierre's Hole"—as Irving calls it—was that of a brave chieftain who there fell by the hands of the Blackfeet.

vous for the fur-hunters and trappers in the early days of western exploration. Some of these "holes" still retain that designation; others are now known as "prairies" or "valleys," the name "hole" having been adjudged not sufficiently elegant. Most of the small grassy areas shut in by the mountains in that portion of the range which is embraced within the Territories of Idaho and Montana are known as "prairies." Farther south, in Colorado and Wyoming, the high plateau-like valleys, which resemble the "holes" and "prairies" of the more northern region, except that they are on a much larger scale, are known as **parks**, and were thus designated many years ago.

The word "park" properly means an enclosure, as does also "paddock" (A. S. *pearruc, pearroc,* a small enclosure) — a word very little used in this country. Skeat says: "It is tolerably certain that *paddock* is a corruption of *parrock,* another form of *park.*" * By

---

* Littré says, of the French word "parc," "mot d'origine obscure," and adds that it is not certain that it comes from the Celtic. It is a word which, in some form, is widely distributed through the languages of Europe. Diez thinks that it is from the Latin *parcere,* to spare or reserve, with the idea that a "park" is a "reservation." This latter word is chiefly used in this country to designate a tract of "park," in this country, is generally understood an enclosure, laid out as a pleasure-ground, planted with a variety of ornamental trees, distributed in picturesque grouping, and rendered accessible by roads and paths. Such a park may be private property, forming a part of a gentleman's estate, but more often a public pleasure-ground, in or near a town or city. When the "parks" of the Rocky Mountains are spoken of, it is usually the more conspicuous ones — the North, Middle, and South Parks — which are intended to be designated. Of these, the North Park is in Wyoming, the others in Colorado. They are areas of various dimensions, walled in by mountains, and lying at a high altitude. The North and Middle Parks are comparatively small, and include but little level land. The South Park is about forty miles long by fifteen or twenty broad, and is more like a plateau or plain than the others. Its northern end lies at an elevation of about 10,000 feet, and it declines toward the south to about 8,000. The San Luis Valley (or Park, as it is sometimes called) lies still farther south, is much larger than the others, and less "park-

land set apart by the Government for some special purpose, as for occupation by the Indian tribes, or for a light-house, or for military defence.

like," being, in fact, in large part a sandy desert where nothing of value can be grown without irrigation. For the other "parks," that name is not entirely fanciful, since they are enclosed by mountains, and in places, especially along the edges of the mountains, ornamented by clumps of trees, which are sometimes very gracefully and picturesquely grouped.

There are several words nearly equivalent in meaning to "plain" and "prairie" which are more or less in use among those writing in English on the physical geography of this and other countries, although they cannot be said to have been adopted into the language, so as to have become "household words," as is the case with "prairie." **Steppe, llano,** and **pampa** are sufficiently familiar, even to school-children, who are early taught that the grassy, treeless plains of Northern Asia and of South America, north and south of the Amazon, are respectively thus designated. Following the example of Humboldt, writers on the physical geography of North or South America not unfrequently speak of the "steppes" of the New World. "Ansichten der Natur" (Aspects, or Views, of Nature) is the title of the well-known and extensively circulated book which has made us so familiar with the word "steppe." In the section of this work entitled "Ueber die Steppen und Wüsten," Humboldt frequently calls the treeless plains of both North and South America "Steppen," the German plural of "Steppe," which is the form the word has in the singular in English, French, and German. It is the Russian степь, a word frequently used in that country in the singular, but by those writing in other languages more commonly put in the plural — the "steppes" (Ger. *Steppen*). The word is defined by Dal as a "treeless, and frequently waterless, uncultivated region of large extent; a desert."* Some additional remarks in regard to the limits and character of the "steppes" will be found farther on, under the word **heath**.

Llano has the same origin and meaning as our "plain," and is used by the Spanish just as we use that word. "Llano" as a noun, and its derivative "llanura" (a llano region), are commonly employed wherever this lan-

* The word here translated "desert" (пустыня) means literally an *empty* region — that is, uninhabited, or thinly inhabited. It is the exact equivalent of the Hungarian *puszta*, which means an empty or uninhabited region. The Hungarian "pusztas" occupy a large portion of the central part of that country, and are given over to cattle-raising. In the "puszta region" there are fertile areas, like oases, where are large estates on which ordinary farming is carried on.

guage is spoken; but the "llanos" proper are the vast treeless plains extending "from the Caracas coast chain to the forest of Guiana, and from the snowy mountains of Merida to the great delta formed by the Orinoco at its mouth" (Humboldt). "This steppe," the author goes on to say, "occupies an area of over a quarter of a million square miles."

The treeless region to the south of the Amazon forest-belt is known as the **pampa**—a name as familiar to us as is that of "llano." "Pampa" is certainly an aboriginal Peruvian (Quichuan) word, although some have sought for its etymological relations in the Latin *pampinus*, from which comes the French *pampre* (the leafy branch or shoot of the grape-vine). But, aside from the fact that there seems to be no connection between a leafy vine and a treeless area, it appears that the word "pampa"—also written "bamba"—is one of frequent occurrence in Peru, where it forms a part of many aboriginal proper names, with the meaning of "level spot," or "field."* The word "pampa" is

* "La falta de llanuras en esta provincia [Pomabamba] es lo que ha dado lugar á que se aplicase la terminacion de *bamba* á todos los lugares donde se encuentra la mas pequeña meseta." (Raimondi, El Perú, vol. i. p. 312.)

not — nor has it ever been — in use in Spain, except as imported (if this expression may be allowed) from South America.

Páramo is another Spanish word much in use in the Andes, but with which we are less familiar than we are with "llano" and "pampa." By "páramo" is generally understood, according to Barcia, "a desert plain, bare of trees, at a high elevation, open to the winds, uncultivated and uninhabited." Humboldt says that the term "páramo" includes all those mountainous regions in the Andes which are from 11,500 to 14,000 feet above the sea-level, and which have disagreeably raw and foggy climate. The vegetation of the Andean páramos is decidedly Alpine, and shrubby or grassy; but this word, as used by some later Spanish writers — Barcia to the contrary, notwithstanding — includes high level tracts covered by dense forests.*

Puna, a word current in the Peruvian Andes, and perhaps in other parts of the chain, seems to be nearly the equivalent of "páramo." Tschudi says that by the name of "puna" is designated the high table-land in

* See Memorias de la Comision del Mapa Geologico de España, Provincia de Cuenca, p. 16, where the author speaks of "altos páramos en donde se desarolla una potente vegetacion forestal," etc.

Peru and Bolivia lying between the two great ranges of the Cordillera, beginning at an elevation of about 10,500 feet above the sea-level, and extending to the region of eternal snow. The highest, wildest, and most desolate portion of the "puna" is called the "puna brava" (wild puna).* Squier, in describing the main chain of the Andes in Peru, says : " Its summit often spreads out in broad undulating plains, or *punas*, varying from fourteen to eighteen thousand feet above the sea, frigid, barren, desolate, and where life is only represented by the hardy vicuña and the condor. This inhospitable region is the great *Despoplado*, or unpeopled region of Peru." † In the Chilian Andes, on the other hand, according to Darwin, it is " the short breathing from the rarefied atmosphere which is called 'puna.'" The same author says, further, that the "puna" is considered a kind of disease, and that he was shown crosses erected over the graves of some who had died "punado." ‡

* Tschudi, Reisen durch Süd-Amerika, Leipsig, 1869, vol. v. p. 197.

† Squier, Peru, Incidents of Travel and Exploration in the Land of the Incas, New York, 1877, p. 9.

‡ Narrative of the Surveying Voyages of the " Adventure " and " Beagle," London, 1839, vol. iii. p. 393.

In some parts of the Chilian Andes the elevated pasture-lands, on which is a scanty growth of grass, but no trees, are designated by the name "talaje." *

The word **campo** (Lat. *campus*, field) is in use, both in Spain and in Brazil, to designate certain tracts resembling our prairies in origin and character. Thus the "Tierra de Campos," near Valladolid, is an old lake-bottom, with an extremely fertile soil, but entirely destitute of forests. The "campos" of Brazil are level or gently undulating tracts in the midst of the dense forest, but themselves nearly or quite treeless. Mr. Bates says of the country around Santarem that it is a "campo region," which he defines as a "slightly elevated and undulating tract of land, wooded only in patches, or with single scattered trees." † Mr. H. H. Smith describes a "campo region" as one having "trees scattered over the surface, not close enough for shade, nor thickly leaved enough to be called luxuriant." ‡

* A. Plagemann, in Petermann's Mittheilungen, Band xxxiii. (1877), p. 74.
† H. W. Bates, The Naturalist on the Amazons, 3d ed., London, p. 176.
‡ H. H. Smith, Brazils, the Amazons, and the Coast, New York, p. 137.

Mr. Bigg-Wither has studied the "campos" more in detail than either of the authors named. He says, among other things: "These little bare patches or *campos* seem altogether out of harmony with the surroundings, not only in their comparative sterility, but also in the configuration of the ground. For whereas, in the forest land surrounding them, it would be difficult to find a level spot of five square yards together, here you may have many square miles of an almost perfect plain; and so flat is it, indeed, in these *campos* that a large proportion of their extent is permanently covered by swamps." * Although Mr. Bigg-Wither does not say so, it would seem almost certain that these level treeless areas are the beds of more or less completely desiccated lakes, and that they are destitute of trees for the same reason that the "campos" of Spain, and many similar tracts in the midst of the dense forests of North America, are; namely, because the material with which these old lake-beds have become filled up is of exceeding fineness — as it must be in consequence of the surrounding conditions — and hence unfitted for the growth of an arboreal vegetation.

* Pioneering in South Brazil, London, 1878, vol. ii. p. 320.

The word **barren**, as an adjective, is in general use wherever English is spoken, as the antithesis of "fertile." A "barren region," a "barren soil," are familiar phrases. The "Barren Grounds" form a well-known feature in the geography of North America, and under this designation are included all the lands in high northern latitudes from the north end of Labrador west to near the base of the Rocky Mountains. As Sir John Richardson remarks: "It is the absence of trees which has given name and character to the 'barren grounds' of North America. The region is low, nearly level and full of lakes, its surface being varied by occasional rocky hills of moderate altitude." *

These "barren grounds" are the exact counterpart of the **tundras** of Northern Europe and Asia, and this term is one well known to students of physical geography. The "tundras" begin in Northern Lapland, although there not designated by this name,† and stretch through Siberia to the base of the chains of mountains bordering the Asiatic continent on the eastern side. Seebohm describes the "tundras" as "naked tracts of slightly un-

* The Polar Regions, Edinburgh, 1861, p. 263.
† Linnæus calls them "terræ damnatæ."

dulating land, rolling prairie or moor, swamp, and bog, full of lakes, and abounding with reindeer moss, on which the reindeer feed." *

The word "barren" is used in various portions of the United States as a substantive, and generally in the plural, and with somewhat different meanings in different regions, but always as including the idea of absence or sparseness of forests. The extensive belt of country running parallel with, but at some distance from, the Atlantic coast, through the Southern States, and covered with a sparse growth of the long-leaved pine (*Pinus palustris*), is known as the "pine-barrens." In some parts of the Mississippi Valley, tracts of country thinly clad with a growth of small or shrubby oaks are sometimes called "oak-barrens," or simply "barrens." This is especially the case in Kentucky. In Wisconsin and some of the other Mississippi Valley States such tracts are known as "oak-openings."

The soil of the Kentucky "barrens" is fertile. Dr. Owen says of this region: "In the early settlement of Kentucky the belt of country over which it [the Subcarboniferous limestone] extended was shunned, and stamped with the appellation of 'Barrens;' this arose,

* Siberia in Europe, London, 1880, p. 55.

in part, from the numerous cherty masses which locally encumbered the ground, in part from the absence of timber from large tracts, and in consequence of the few trees which have here and there sprung up, being altogether a stunted growth of black-jack oak, *quercus ferruginea*, red oak, *quercus rubra*, and white oak, *quercus alba*." As soon as we pass from the limestone on to the conglomerate, a rock furnishing by its disaggregation a coarser material than that which is left by the decomposition of the limestone, we come upon a densely forested region.*

The "barrens" of Newfoundland are described by Jukes as being "those districts which occupy the summits of the hills and ridges and other elevated and exposed tracts. They are covered with a thin and scrubby vegetation, consisting of berry-bearing plants and dwarf bushes of various species, and are somewhat similar in appearance to the moorlands of the North of England, differing only in the kind of vegetation, and in there being less of it." † In other portions of Northeastern

* See Kentucky Geological Survey Reports, New Series, vol. i. p. 32, where this fact is admitted, but its theoretical importance overlooked, and elsewhere in the Report denied.

† J. B. Jukes, Report of the Geological Survey of Newfoundland, London, 1843, p. 22.

Canada treeless areas, from whatever cause originating, are called "barrens." Some of these, as described by various observers, are unquestionably the result of the desiccation and gradual filling up of lakes, and treeless areas of this kind are found all through the region of the Great Lakes, surrounded by the densest forests; but the use of the word "barren" as designating them is — so far as the present writer has observed — limited to the extreme northeastern portion of the great Atlantic forest-belt.

In the densely wooded portion of the United States, the first thing the settler has to do is to cut down the trees over an area of sufficient size for cultivation and for the necessary buildings. The piece of ground thus prepared is called a **clearing**, a household word in the newly settled forested regions.* An **opening**, on the other hand, is a natural deficiency of the forests over a certain area, the trees not being entirely wanting, but thinly scattered over the surface as com-

---

* The word "clearing" is universally used by the German immigrants, instead of "Lichtung," to designate the locality thus prepared for a "settlement." The present writer has often heard the occupied land on the frontier spoken of by Germans as being already "gecleared and gesettled."

pared with their abundance in the adjacent region.

A **glade** is also an "opening" or "clearing" in the forest, and this word may be applied either to a space naturally destitute of trees, or to one where they have been removed by the hand of man.* "Glade," however, is a word not much in use in this country, although there are regions — as, for instance, in Kentucky — where it is current. In that State the phrase "glady ground" is sometimes heard, and by it is meant a district where the surface is diversified by alternate forests and openings. In parts of Virginia and the adjacent States, localities in the "timber" which are too wet for a forest growth, but which are more or less overgrown by bushes, are designated as **slashes**. This word is said to be applied also in the Northern States to the tracts covered with fallen timber left by the passage of a tornado through the forests, and also to land "on which the underbrush has been cut and left lying." † "Slash," as a

---

* "Glade" is defined by Skeat as "an open space in a wood." Wedgwood says: "a light passage made through a wood, also a beam or breaking in of the light." It is a word of Scandinavian origin; the original sense being an opening for light, a bright track, hence an open track in a wood.

† Bartlett, Dictionary of Americanisms.

topographical designation, seems to be decidedly an Americanism. "To slash," "to cut with a violent sweep, to cut at random" (Skeat), is a word in common use both in England and in America.

"Glade" is a favorite word with the poets, especially with Scott, who seems to delight in its use. In the works of American authors it is much less frequently found. There is a certain vagueness of meaning in the word "glade" which makes it very convenient for poetical use, as may be seen in the following examples: —

> "A forest glade, which varying still,
> Here gave a view of dale and hill,
> There narrower closed, till overhead
> A vaulted screen the branches made."
> *Scott, Marmion.*

> "Here rise no cliffs the vale to shade;
> But skirting every sunny glade,
> In fair variety of green
> The woodland lends its sylvan screen."
> *Rokeby.*

> "Lovely between the moonbeams fell,
> On lawn and hillock, glade and dell."
> *Lord of the Isles.*

> "Thy bounteous forehead was not fann'd
> With breezes from our oaken glades."
> *Tennyson, Eleanore.*

"The mossy bank, dim glade, and dizzy height."
                                            W. S. Landor.

"Here, say old men, the Indian magi made
Their spells by moonlight; or beneath the shade
That shrouds sequestered rock, or darkening glade,
    Or tangled dell."
                                    J. G. C. Brainard.

The grassy summits of various high mountains in the extreme southern extension of the Appalachians are known as **balds**. Mr. J. W. Chickering thus describes one of them: "The top [of Roan Mountain], instead of being, as in the higher of our New England peaks, a mass of barren rock, or weather-worn boulders for the upper 1,000 feet, is a smooth grassy slope of 1,000 acres, called a 'bald' (the soil a foot or more deep and as rich and black as a western prairie), with rocky precipices at either end, rising 80 to 100 feet higher, but plentifully covered with [forest?] vegetation." *

The words "fell," "wold," "moor," "down," and "heath" are familiar to all who read about the geography and history of the British Islands; but, as actual designations of the features of the landscape, they are almost entirely unknown in the United States. It is not easy to draw a sharp line of distinction between these words, and no one of them is

* Appalachia, vol. ii. p. 277.

strictly limited to any particular part of Great Britain, although there are certain regions where each is, in a measure, localized. Thus, fell is a topographical designation but little known outside of the Lake District and its immediate vicinity. It is the Sw. *fjäll*, Ice. *fjall*, *fell*, Nor. *fjeld*, M. E. *fel*, and essentially the same (etymologically) as the Eng. *field*. In Norway the word "fjeld" (pl. *fjelden*) is one in general use to designate the high, table-topped mountains which are so conspicuous a feature in the physical geography of that country. Professor Forbes calls them "those wonderful expansions of mountains, often so level, that upon what might almost be called their *summits* a coach and four might be driven along or across them for many miles, did roads exist, and across which the eye wanders for immense distances, overlooking entirely the valleys, which are concealed by their narrowness, or by small mountains which rise here and there with comparatively little picturesque effect above the general level." *

In the Lake District the word "fell" is used with essentially the same meaning as

* J. D. Forbes, Norway and its Glaciers, London, 1853, p. 191.

that given above, except, of course, that the mountains of that region are not so grand and high, and not so distinctly "table-topped," as they are in Norway. Black's Guide to the Lakes gives the word "fell" as meaning "bare, elevated land, and answering in some respects to the wolds, moors, and downs of other parts of the island." It is a term in common use in every part of the Lake District, where, however, this designation is by no means exclusively limited to high table-lands or flat-topped hills, but is occasionally used for any rocky eminence. Skeat, indeed, defines the word "fell" simply as "hill." Black's Guide gives the following quotation from an old manuscript: —

> "Moyses went up on that felle,
> Fourty dayes there you dwell."

The word "fell" seems never to have obtained a foothold in this country. An attempt has been made to localize it here, however, by giving the name of "Middlesex Fells" to a rough, rocky district a few miles north of Boston, in which lies a pretty lake known as "Spot Pond." It is not easy to see any particular appropriateness in the name "fell" as applied to this locality, which does not differ essentially from the ordinary type of New England landscape.

The word **moor** (Ice., Dan., A. S. *mor;* M. E. *more;* Ger. *Moor*) is less easy to define than "fell," as it is used with a variety of significations in different parts of Great Britain. It is curious to see how the dictionaries differ in attempting to indicate its meaning. It is decidedly most familiar to us as designating those tracts in Scotland on which game is preserved, and to which fashionable people resort in the autumnal season to while away the time and enjoy the high privilege of killing something. The Scotch "moors" are the elevated, undulating, treeless, flat or gently sloping tracts, from which rise the ranges of precipitous hills and mountains which characterize the grand but at the same time rather gloomy and monotonous scenery of the Scottish Highlands. They are "fells," but fells crowned with still higher and more precipitous summits, which themselves are sometimes fell-like in character. Thus, Geikie describes the mountains at the head of Glen Esk and Glen Isla as sweeping upward into a broad "moor" some 3,000 feet above the sea; in regard to which he remarks that it would hardly be an exaggeration to say that there is more level ground on the tops of these mountains than in areas of corresponding size in the valleys below.*

In the more southern portions of Great Britain "moor" seems to be used nearly as the equivalent of **morass**, which latter word is said by Skeat to be plainly an adjectival form of "moor," and is defined by him as "swamp, bog;" while Latham defines "moor" as "marsh, fen, bog, tract of low and watery ground."

The famous "moor"—"the great central waste of Devon"—named from the river Dart (Dartmoor), is, however, far from being a "morass." Mr. A. N. Worth thus indicates its peculiar topographical and geological features: "In the main it is a great granitic plateau, broken by numerous valleys, and dotted with the rocky peaks of the 'tors.' The granite is jointed, often with considerable regularity, and weathers into masses and piles irresistibly suggestive of Cyclopean masonry; while the hillsides are bestrewn for miles with huge boulders and blocks." † All Devon, according to Mr. Worth, was formerly charac-

* A. Geikie, The Scenery of Scotland, 2d ed., London, 1887, p. 195.
† A History of Devonshire, London, 1886, p. 330.

terized by woods and heaths, broken only in their gloomy monotony by strips of water-made meadow skirting the wider river-courses, and the scanty population was scattered indifferently through its wilds. "Dartmoor is simply the last refuge of the traces of these ancient days — a prehistoric island, girdled and wasted by the encroaching waves of an aggressive civilization." *

The present writer never heard the word "moor" used in this country as designating any feature of our landscape; but "morass" is occasionally used, although this latter term is rather the elegant designation of what is popularly known as a **swamp** — a word current all over the United States with the meaning of low, marshy ground, whether thickly or thinly forested, entirely bare of trees, or covered with a shrubby vegetation. Thus a low, flat piece of ground, on which the water stands during a portion of the year, and over which is a sparsely scattered growth of tamarack or hackmatack (*Larix Americana*), is familiarly known as a "tamarack swamp." To the "cedar swamps," so characteristic of the Upper Lake Region, allusion has already been made.† A peculiar swampy region

* A History of Devonshire, London, 1886, p. 326.
† See *ante*, p. 186.

extends along the Atlantic coast, from Virginia through North and South Carolina, of which the "Great Dismal Swamp" on the borders of Virginia and North Carolina may be taken as the type. These swamps have certain peculiar features, the most important of these being that they are considerably elevated above the adjacent streams, and their forest vegetation is abundant and varied, the most characteristic tree being the cypress (*Taxodium distichum*). These swamps are locally known as "dismals" and also as "pocosins," the latter being apparently an aboriginal name, and, if so, one of the very few instances (if not the only one) in which a word of this kind has become — to a limited extent, it is true — generalized as a topographical designation.

Skeat defines the word "swamp" as "wet, spongy land, boggy ground," and adds "not found in old books."* He considers it as being of Scandinavian origin (Dan. and Sw. *svamp*, a sponge, fungus; Ger. *Schwamm*, a sponge), and remarks that "swamp," "sponge," and "fungus" are all related words, and all from the root of "swim." "Swamp" seems peculiarly an American

* Wafer uses the words "swamp" and "swampy" frequently in his "New Voyage and Description of the Isthmus of America," London, 1699.

word; and the so-called "swamp lands" forming a portion of the national domain have been freely bestowed on the various States in which they occur, and have been the source of endless fraud and deceit, since large areas of the most valuable agricultural land in the country have been claimed and held as "swamp land."

Swale is a word not to be found with a topographical meaning in English dictionaries, but frequently heard in the United States, especially in the Prairie States, where it is used to designate the depressions, or lower, moister areas, in the "rolling prairie." The definition of "swale" given in Webster — "an interval or vale; a tract of low land" — does not agree with the present writer's experience of the use of the words "swale" and "interval." A "swale" is always a lower area of moderate dimensions, in the midst of higher ground, and it would never be used as the equivalent of "interval" or "intervale." * "Swale" is a word current in East Anglia, meaning there, according to Nall, just what it does in this country, "a low place, a hollow." It comes from the Scandinavian (Dan. *svælg*, a hollow, an abyss).

* See pp. 228, 229, for definition of "interval" and "intervale."

The word "dun" (Gael. *dun*, Welsh, *din*, A. S. *dún*, a hill) appears in England and elsewhere in several rather peculiar and interesting forms. In the Lake District **dun** means an inconspicuous hill; and with this signification it forms a part of certain proper names, — for example, Dunmallet, Dunfell, etc. Farther south, this word has the form of **down**, and is used to designate various elevated, flat or gently undulating, treeless areas, underlain by the chalk, and mostly given over to sheep-raising. The "Downs" are a peculiar feature of English scenery, and "South Down mutton" is a term which needs no explanation. The Downs proper are in Kent and Sussex, those in the first-named county being called the North, and the other the South Downs. They lie on each side of the curious depression known as the "Valley of the Weald," or "Wealden," or simply as "the Weald." Other areas of similar geological and topographical character are also called "downs." The word is a favorite with the poets, and especially with Tennyson, who by no means limits it to England, but, on the contrary, puts downs and palms together, as may be seen in the following extract from the "Lotos Eaters:" —

> "And the yellow down
> Border'd with palm, and many a winding vale."

The use of the word "down," as a topographical designation, is almost, if not quite, unknown in the United States. The present writer has been able to find but one poem by an American writer, on American scenery, where it is introduced: —

> "With music that rises and falls and swells,
> Over the village and past the down,
> Past Katama and Roaring-Brook,
> Out by Gay Head, where, at set of sun,
> The light-house gleams over hill and nook."
> *E. N. Gunnison, The Bells of Edgartown.*

That part of the English coast adjacent to the region where the North Downs meet the sea is also known as "the Downs."

The word "down" has found its way to Australia, where "the Darling Downs" is the name of a district lying west of Brisbane, in Queensland, and the seat of the most important agricultural interests of that colony. As described by Australian authors, this region is "mainly a huge plain, where the surface, which sometimes rises into rolling downs and sometimes spreads out in apparently limitless flats, is only broken by a few ranges of low hills." *

---

* C. A. Feilberg, in "Australian Pictures," p. 117.

There is still another form of the word "dun," by which certain hills are designated, but only those of a peculiar origin and character. Hills of loose sand, heaped up and blown about by the wind, especially along the sea-coast, are called **dunes**, a word which has the same form and meaning in French, and nearly the same in German (*Düne*). Dunes occur chiefly along the sea-shore, but are also seen on the borders of large lakes, for instance Lake Superior, along portions of whose southern shore they rise to the very respectable height of fifty feet or more. Moving sands in the interior, as in various desert regions, are also sometimes designated as "dunes"—more frequently, however, as "sand-hills." On the East Anglian coast the sand-dunes are called **meals**, a relic of the Norsemen (Ice. *möl*, strand sands) (Nall).

Wold is another word quite peculiar to England, and of some obscurity, both as to meaning and origin. As used in parts of Yorkshire, it seems to be the exact equivalent of the "downs" of Southern England. The "Wolds" of that county form a crescentic range of elevations, sloping from a curved summit, whose extremities touch the sea at Flamborough Head and the Humber at Ferriby, and this crescent is cut through by one continuous hollow — the Great Wold Valley — from Settrington to Bridglington. As is the case with the "downs," so here the underlying formation is the chalk. The same word appears in the form of "weald," a term especially familiar to geologists, as having given the name to the "Wealden formation," which occupies the basin-like depression between the chalk escarpments of the North and South Downs, and to which reference has already been made.

The term "wold" would seem from its general application in England to be intended to designate an open, unforested region. It is, in fact, defined by Latham as a "plain, open country;" and by Skeat as a "down, open country." It is a favorite word with the English poets, who sometimes use it rather vaguely, but more generally with the meaning given above, as the following quotations seem to indicate : —

> "Long fields of barley and of rye,
> That clothe the wold, and meet the sky."
> *Tennyson, The Lady of Shalott.*

> "Arise and let us wander forth,
> To yon old mill across the wolds."
> *Tennyson, The Miller's Daughter.*

The following is the only instance (so far as known to the present writer) in which "wold" has been used by an American poet : —

> "Never errant knight of old,
> Lost in woodland or on wold,
> Such a winding path pursued
> Through the sylvan solitude."
> *Longfellow, The Songo River.*

The word "wold" is considered by some etymologists as being the German *Wald*, (M. E. also *wald*); which, however, as Skeat remarks, was more commonly used in the sense of "waste ground, wide open country," as in Norse, and this statement is substantiated by authorities cited by him. He adds as follows : "The connection in form with A. S. *geweald*, Ice. *vald*, dominion, is so obvious that it is difficult to assign any other origin than Teut. *wald*, to rule, possess, for which see *wield*. The original sense may have been 'hunting-ground,' considered as the possession of a tribe." Some writers have argued that because certain regions are known as "wolds," they must originally have been forested; but this seems decidedly improbable, in view of the fact that the areas thus designated have precisely that geological character which is unfavorable to the growth of trees, as is the case both with the "wolds" and the "downs," which — so far as historical evidence goes — have always been what they now are, namely, open treeless regions, and for which condition there is abundant reason to be found in the peculiar fineness of the soil, resulting from the decomposition and decay of the chalk, as well as of a large portion of the various strata, of which the Wealden group of the geologists is made up.*

Closely allied to "down," "wold," and "moor," is the word **heath**, which is much used in England and Scotland, but quite unfamiliar to us except through books. "Heath" is defined by Skeat as "a wild, open country;" by Latham, as "a place overgrown with heath," or "a place covered by shrubs of any kind." Absence of forests seems the essential feature of a heath; nor is it easy to see, either from dictionaries or from other books or from its poetical use, in what a "heath" differs from a "moor." Indeed, Skeat defines a "moor" as

---

* The whole of the Wealden area is not treeless, but the larger portion of it is so; and the soil of this portion has been described by competent authority as being, when dry, "an impalpable silicious dust."

a "heath." It is etymologically the same as the German "Heide" (M. E. *heth* and *hethe*, Swed. *hed*, Dan. *hede*, Du. *heide*).* The "Heiden" of North Germany are an important topographical feature of that country, and they pass gradually into the "steppes" of European and Asiatic Russia; for the great "steppe region" begins on the very borders of Holland, and extends in unbroken continuance, save where interrupted in part by the chain of the Ural, almost to the farthest eastern limits of Siberia. The vegetation of the "heaths" is somewhat varied; but by far the most characteristic heath plant is that called "heather" (heath-er, inhabitant of the heath), or also frequently simply "heath." Humboldt, in the chapter of the Aspects of Nature entitled "Physiognomy of Plants," says: "The Heath form belongs more especially to the Old World, and particularly to the African continent and islands. . . . In the countries adjoining the Baltic, and farther to the north, the aspect of this form of plants is unwelcome, as announcing sterility. Our heaths, Erica (Calluna) vulgaris, Erica tetralix, E. carnea, and E. cinerea, are social plants,

* "All from an Aryan base KAITA, signifying a pasture, heath, and perhaps clear space." (Skeat.)

and for centuries agricultural nations have combated their advance with little success. It is remarkable this extensive genus which is the leading representative of this form appears to be almost limited to one side of our planet. Of the 300 known species of Erica, only one has been discovered across the whole extent of the New Continent, from Pennsylvania and Labrador to Nootka and Alashka."* The common heather (*Calluna vulgaris*) has been found in various localities along the Atlantic coast, from Massachusetts to Newfoundland; but the word "heath" as a topographical designation appears to be quite unknown in this country. Near London there are various tracts denominated "heaths," the names of which are very familiar to readers of English plays and novels. Most of these "heaths" are outliers of the Bagshot Sands; and where these attain

* Aspects of Nature (Mrs. Sabine's translation), vol. ii. pp. 23, 24. In a note to this the author adds: "In these physiognomic considerations we by no means comprise under the name of Heaths the whole of the natural family of Ericaceæ, which, on account of the similarity and analogy of the floral parts, includes Rhododendron, Befaria, Gaultheria, Escallonia, etc. We confine ourselves to the highly accordant and characteristic form of the species of Erica, including Calluna (Erica) Vulgaris, L., the common heather."

their full development — that is, where the formation retains its entire thickness of 300 to 400 feet — the depth to the water-level becomes so great that the upper porous beds are left high and dry, and form uncultivated wastes, such as Bagshot Heath, Frimley Heath, and others. These are still for the most part bare "heaths," which, being sandy, dry, and healthy, have been frequently used for military camps and exercise-grounds.

With the English, and still more with the Scottish poets, "heath," "heather," "heath-bells," are favorite words; not much less so are "bracken," "gorse," and "broom" — other characteristic shrubs which help adorn the heaths. A few quotations may be added as illustrations of the poetical use of these words: —

> "But most, with mantles folded round,
> Were couch'd to rest upon the ground,
> Scarce to be known by curious eye,
> From the deep heather where they lie,
> So well was matched the tartan screen
> With heath-bell dark, and brackens green."
> *Scott, Lady of the Lake.*

> "The great fires are luntin' — how fragrant the smells,
> The bab o' the heather, and bonnie bluebells,
> This twig o' green birk — oh, I canna weel tell
> Hoo the sicht and the scent gars my fu' bosom swell."
> *Janet Hamilton.*

There is in France a region adjacent to the ocean, north of the Pyrenees, which in some respects resembles the "heaths" of Northern Germany. It was once the bed of the ocean, and is covered with sands of Pliocene age. The natural growth of these **landes**, as they are called, consists of heather, broom, and ferns, much resembling that of the more northern "heaths." Over some portion of these "landes" the introduction of a forest growth has been successfully attempted; in other districts the presence near the surface of a solidly compacted bed of sand has proved an insurmountable obstacle to tree-culture.

Marsh is a word in common use in both England and the United States, with a meaning not essentially different from that of "swamp." Skeat defines it as "morass, swamp, fen," and says that it has the form in Middle English of "mersche," and in Anglo-Saxon of "mersc," which latter is a contraction of "mer-isc," originally an adjective signifying "full of meres or pools." As used in the United States, the word "marsh," often pronounced "ma'sh," is heard much more frequently along the sea-coast in New England than it is in the interior and farther south. The low lands along the New England coast liable to overflow by the tide are always called "salt-marshes." The same word "marsh" is one commonly used in Northern Germany in almost exactly the same way in which it is used in this country. The "Marschländer" (called also simply the "Marsch") form an important topographical feature along the coasts of the Baltic and North Seas, and especially in the vicinity of the Elbe. They are uniformly level, the monotony of the surface being hardly broken by the dikes by which they are traversed at regular intervals, and the ditches which accompany them. On the dikes grow magnificent trees, the soil is very fertile, the cattle superb, and farming highly successful. The contrast between these marsh lands and the region of sand and gravel — the "heath" and "moorland" — which lies adjacent to them on the south is most striking.*

Moss is a word very familiar to us as the name of an order of the class of Cryptogams — the *Musci*, or Mosses. In Northern England and Scotland it is also much used to designate various localities which are swampy or boggy in character, and especially those

* See E. H. Wichmann in Zeitschrift der Gesellschaft für Erdkunde zu Berlin, vol. xx. (1885) pp. 257-279.

where peat is found in some quantity. In Southwestern Yorkshire the peaty mountains are called "mosses." In Scotland we hear of "moss-troopers"—a name formerly given to those horsemen who rode over the high, peaty moorlands. Thus, Scott says:—

> " A stark moss-trooping Scott was he,
> As e'er couched Border lance by knee;
> Through Solway sands, through Tarras moss,
> Blindfold he knew the path to cross."
> *Lay of the Last Minstrel.*

> " He journey'd like errant knight the while,
> And sweetly the summer sun did smile
> On mountain, moss, and moor."
> *The Bridal of Triermain.*

What are called in New England "peat-swamps" or "peat-bogs," are known in Northern England and Scotland as "peat-mosses." They may be, as in Yorkshire, at a high elevation. In Lancashire low, boggy places are called "mosses;" for instance, Carrington Moss, and Chat Moss; the latter the locality famous for having presented such extraordinary engineering difficulties in the course of the construction of the first surface railroad. In this part of England, where the low, swampy grounds are called "mosses," the highlands are designated "fells" and "moors." **Bog,** a word of Irish origin (*bogach*, a morass), is often heard in this country, and seems to be the exact equivalent of "morass." "Mire" (Dan. *myr, myre,* Swed. *myra,* M. E. *myre,* O. H. G. *Mios,* M. H. G. *Mies,* moss, morass, swamp), is etymologically related to "moss" and "morass," but is hardly to be considered a topographical word. Like "mud," it means the material which helps fill up miry, boggy, or muddy localities — not only those on a large scale, like morasses and swamps, but smaller ones, such as roads, ditches, and hollows generally.

The word **fen** (A. S. *fen,* Du. *veen,* Goth. *fani,* Ger. *Fehn*) is defined in the English dictionaries as the equivalent of "morass" or "bog." It seems, however, as actually used, to mean ground wet enough to be more or less thickly overgrown with reeds and other aquatic vegetation. The "Fen District" of England is a wide stretch of level, monotonous marsh, traversed by a multitude of sluggish streams, situated within the counties of Lincoln, Cambridge, Norfolk, Suffolk, Huntingdon, and Northampton, and extending about fifty miles from north to south, and thirty from east to west. It was in former times a swampy, unhealthy wilderness; but it has been drained at immense cost of money and labor, and made one of the most fertile portions of the kingdom. A portion of this fenny district is known as the **broads**, a term peculiar to Norfolk; and the relation of the "broads" to the "fens" is easily understood. Where the rivers broaden out and are more or less separated into distinct channels by belts of reedy growth (reed-beds), there "we find as the result a region where water and land strive for the mastery and come to a delightful compromise." * The author from whom this is quoted adds further: "The character of the Fens has been so much changed since their drainage, that it is to Norfolk only that one can now look for the wildness and solitude of marsh and mere so dear to the naturalist and sportsman."

" Fen " is a word little known in this country except through books and in poetry. The low, swampy tracts of country in Florida known as "the Everglades" are something nearly akin to "fens." Mr. Barbour thus describes this region: " Perhaps the most remarkably geographical feature of the State [of Florida] is the immense tract of marsh or lake, called the Everglades (by the Indians " grass-water ").

* G. Christopher Davies, Norfolk Broads and Rivers, Edinburgh, 1883, p. 2.

It is about sixty miles long by sixty broad, covering most of the territory south of Lake Okechobee, and is impassable during the rainy season, from July to October. The islands with which its surface are studded vary from one-fourth of an acre to hundreds of acres in extent, and are usually entangled in dense thickets of shrubbery or vines. The water of the lake is from one to six feet deep, and the bottom is covered with a growth of rank grass which, rising above the surface, gives it the deceptive appearance of a boundless prairie." *

Scrub and **scrogg** are closely allied to each other, both in origin and meaning; and both are botanico-topographical designations in various regions where English is spoken. They signify land covered with a stunted, scraggly, or shrubby undergrowth. " Shrub " is a word in common use with us, as in England, meaning something midway between a tree and an herbaceous plant. It is nearly the equivalent of " bush ; " and a " shrubbery " is a place covered or planted with shrubs, although the use of this word is pretty closely limited to an artificial plantation or garden of shrubs. The natural growth of a region covered with bushes

* Florida for Tourists, Invalids, and Settlers, p. 20.

would hardly be called — in this country, at least — "shrubbery," but rather "undergrowth," "underbrush," or simply "brush," "brushwood," or (more rarely) "bush."

Bush, as a topographical designation, is a word more used in the English colonies than with us. Thus, Mrs. Moodie's pleasant narrative of her experience in Canada is entitled "Roughing it in the Bush." The lawless vagabonds who roamed over the uncultivated districts of Australia (the "scrub") were, in former times, generally designated as "bush-rangers," while the natives of South Africa are known as "Bushmen." *

"Shrub" and "scrub" are referred back to the Anglo-Saxon "scrob" (M. E. *shrob, schrub*), and "scrog" is a provincial form of "scrub." The verb "to scrub," a word in familiar use, means "to clean or scour with a bunch of twigs or shrubs," just as we say "to brush," which originally meant "to clean with an implement made of brushwood (twigs or

* The word "bushman" is also used in Australia with the same meaning which "woodsman" has in the United States — namely, as designating a man familiar with the forest, and well able to take care of himself in a "wild country." A "backwoodsman," on the other hand, is rather one who has taken up his abode on the frontier, or far from the settlements.

small branches). The use of "scrub" as a topographical designation is hardly more known in England than it is with us, but some names of localities — for example, Wormwood Scrubbs — show that this word is not altogether strange in that country. In Australia it seems to be a very familiar term, the forest undergrowth being generally denominated "the scrub," while in some parts of that country "scrub" and "bush" seem to be almost equivalent words, although the latter is more often used as meaning both forest and scrub, or any kind of uncultivated or uncleared land, as distinguished from that which has been brought under cultivation.*

The word **scrogg** seems to be limited in use to the North of England. Thus, Gawin Douglas, the Scotch poet (1474–1522), in describing a morning in June, says: —

" And schortlie, everything that dois repare
In firth or feyld, flude, forest, earth, or ayr,
Or in the scroggis, or the buskis [bushes] ronk,
Lakis, marrasis [morasses], or their pulis [pools] donk,
Astabillit [enstabled] liggis still to slepe, and restis."

* Thus the author of "Australian Pictures" (Howard Willoughby) says: "There is something very solemn in the quietude of a scrub untouched by the axe of the lumberer or settler. There is no undergrowth, properly speaking, though delicate little ferns and fairy-like mosses nestle close to the feet of the trees."

In parts of North Germany the shrubby meadow-land is called a "Brül" or "Brühl," a word defined in Grimm as "pratum palustre," or "buschigte Wiese" (bushy meadow). "Breuil" and "broussailles" in French ("brosse" in Old French), and "Brül" or "Brühl" in German, have the same original meaning as our "brush" (or "brushwood"), a word with which they are etymologically connected. "Egerde," or "Egert," is also a name given in various parts of Germany to barren, uncultivated fields, more or less covered with heath and shrubby vegetation. The proper meaning is said by Grimm to be "fallow" or "fallow-land" (Ger. *Brachland*), but the origin of the word is obscure.*

Coppice, copse, and **coppy** are words frequently heard in England, and with which we are very familiar through English books. They can, however, hardly be said to form a part of our vocabulary. "Copse" and "coppy" (the latter not nearly as often met with in print as the former) are variants of "coppice," which is from the Low Latin *copecia*, undergrowth, and is allied to the

---

* Grimm says "denkbar wäre âgartia, âgertia, âgerta, ungesäumtes, ungehegtes, der weide preis gegebenes ackerland."

French *couper*, to cut. Any area covered with a shrubby undergrowth may be called a "coppice;" but, as the word is generally used, it means a plot of ground where such an undergrowth is maintained and kept down by being frequently cut for fuel. The twigs thus obtained are made up into bundles called "faggots" — a word also very familiar to us through English books, but rarely, if ever, heard in actual use.

"Interval" and "bottom," as topographical designations, appear to be peculiarly American words. An **interval** (Lat. *intervallum*) is the space between a river and the hills or mountains by which the lower, level portion of the river-valley is bounded. Hence "interval" has nearly the same meaning as "meadow," and the two words are more or less interchangeable; the level, cultivated, and frequently grassed areas bordering the Connecticut River, for instance, being generally called "meadows" or collectively "the meadows." **Intervale** is a variant of "interval," less frequently used than the latter word. Some villages on or near tracts of interval land are called by the name "Intervale," as, for instance, the summer resort thus designated in the valley of the Saco River, near North

Conway, in which region the word "intervale" seems to be much more frequently used than "interval." Thus, Whittier says : —

"From the heart of Waumbek Methna, from the lake that never fails,
Falls the Saco in the green lap of Conway's intervales."

And an anonymous author, describing the Kennebec, has as follows : —

"You look upon a range of intervales
Where the abundant harvest never fails."

Bottom is a word frequently heard in the Mississippi Valley and farther west, and used to designate the alluvial tracts along the river-courses, which are sometimes called "bottom-lands" and sometimes simply "bottoms." Josselyn (1675) uses the word thus : "swamps, which are low grounds and bottoms infinitely thick set with Trees and Bushes of all sorts." Lewis and Clarke also frequently employ the word "bottom" in their report, and speak of the "American Bottom," an extensive tract of level and highly fertile land stretching along the Mississippi River southward from the Kaskaskia River for many miles.

There are some words locally used in various parts of Great Britain, to designate the alluvial lands, or meadows, bordering the rivers.

Thus, at Bath on the Avon, there are the "Dolly Meadows," "dolly" being the Welsh *dolau*, a meadow, this being one of those reduplications in names which so frequently occur, and various instances of which have already come under our notice. In Scotland the meadows along the streams are called "haughs," a word allied to "haw" and "hedge," having the original meaning of "enclosure." The level tracts of alluvial lands bordering the estuaries along the coast of Scotland are known as "carses." They are marine terraces, or old beaches which have been raised to varying elevations above their former position. One of these "raised beaches" is thus described by Geikie: "The twenty-five feet beach must be more or less familiar to every one who has visited almost any part of the coast-line of Scotland. It runs as a terrace along the margin of the Firth of Forth; it forms the broad Carse of Gowrie; it is visible in sheltered bays along the storm-swept coasts of Forfar, Kincardine, and Aberdeen. In the less exposed parts of the Moray Firth it may be traced, and westwards around most of the northern firths it runs as a conspicuous feature. On the Atlantic side of the island, its low green platform borders both sides of the Firth of Clyde, fringes the islands, runs up the river beyond Glasgow, and winds southwards along the coast of Ayrshire and Wigton into the Irish Channel." *

* Scenery of Scotland, pp. 382, 383.

INDEX

OF TOPOGRAPHICAL NAMES.

ABYSM, 157.
Abyss, 157.
Aiguille, 118.
Amphithéâtre, 170.
Atravieso, 140.

BALANCED-ROCK, 127.
Bald, 205.
Ballon, 114.
Band, 103.
Barf, 104.
Barrancal, 151.
Barranco, 151.
Barren, 199–201.
Barren Grounds, 199.
Bay, 81.
Belchen, 114.
Ben, 102.
Blow-hole, 172.
Bluff, 104.
Bog, 225.
Boiler, 172.
Bölchen, 114.
Boquete, 140.

Bottom, 232.
Box-cañon, 153.
Brèche, 139.
Broad, 226.
Broussailles, 230.
Brühl, 230.
Brül, 230.
Buller, 172.
Bush, 228.
Butte, 105.

CADENA, 87.
Caire, 102.
Cajon, 132, 153.
Camel's Hump, 120.
Campo, 197, 198.
Cañada, 132, 133.
Cañon, 132, 133.
Carcabucho, 153.
Cárcova, 152.
Cárcovo, 153.
Carse, 233.
Catena, 87.
Cau, 102.

INDEX.

Cerrito, 100.
Cerro, 100.
Chain, 85, 87.
Chaine, 87.
Chasm, 155, 156.
Cirque, 170.
Clearing, 202.
Cleugh, 169.
Cliff, 121.
Clough, 169.
Clove, 169, 170.
Cluse, 168.
Cobble, 109.
Col, 136.
Colina, 100.
Collado, 100, 140.
Combe, 112, 164–168.
Coom, 164, 165.
Coppice, 230, 231.
Coppy, 230.
Copse, 230.
Cordillera, 87–90, 99.
Cordon, 87, 99.
Corry, 165, 166.
Coste, 102.
Cove, 81, 163.
Crag, 122.
Crest, 112.
Crest-height, 134.
Creux, 170, 171.
Cwm, 164.

Dale, 142.
Dalle, 143, 144.
Dean, 144.
Debris-pile, 125.
Defile, 148.

Dell, 142, 143.
Den, 144, 145.
Dene, 144.
Dent, 116.
Devil, 163.
Dingle, 145, 146.
Dimble, 145, 146.
Divide, 141.
Dodd, 108.
Doldc, 108.
Dome, 114.
Door, 137.
Dore, 137.
Down, 213, 214.
Druid-stone, 128.
Dun, 213.
Dune, 215.

Egerde, 230.
Egert, 230.
Escarpment, 124.
Espigon, 97, 100.
Esquerra, 101.
Everglade, 226.
Ezquerra, 101.

Fell, 205–207.
Fen, 225.
Fjeld, 206.
Flume, 158, 159.

Gap, 135.
Garganta, 152.
Ghyll, 145, 146.
Gill, 145, 146.
Glade, 203, 204.
Glen, 147, 148.

Gorge, 149, 150.
Gouffre, 157.
Gray-wether, 128.
Group, 85.
Gulch, 154, 155.
Gulf, 81, 156–158.
Gully, 154.

HARBOR, 81.
Hause, 137.
Haws, 137.
Hay-stack, 119.
Head, 111.
Heath, 218–221.
Hêche, 102.
Hell, 162.
Highland, 182.
Hog-back, 119.
Hole, 81, 189.
Hollow, 135.
Hope, 169.
Horn, 115.
Horse-back, 120.

ICE-GULF, 158.
Interval, 231.
Intervale, 231.

JOCH, 140.

KAMM, 112.
Kette, 87.
Kettle, 171.
Knob, 106.
Knock, 107.
Knoll, 107.
Knot, 107.

Kofel, 110.
Kogel, 111.
Kopf, 110.

LANDE, 222.
Llano, 192, 193.
Logan, 127.
Loma, 101.
Lomería, 101.
Lomita, 101.

MARSH, 222, 223.
Meal, 215.
Mesa, 182.
Meseta, 182.
Montaña, 99.
Montañuelo, 99.
Monte, 99.
Monument, 126.
Moor, 208–210.
Morass, 209.
Moss, 222, 223.
Mound, 109.
Mount, 91, 102.
Mountain, 91.

NECK, 136.
Needle, 117.
Notch, 135.

OAK BARREN, 200.
Ocean, 80.
Olla, 170.
Opening, 202.

PAMPA, 192, 194.
Pap, 118.

Páramo, 195.
Park, 190, 191.
Parks, The, 191.
Paso, 136.
Pass, 135.
Passe, 136.
Pass-height, 134.
Peak, 93–95.
Pech, 97.
Pen, 102.
Peña, 97, 98, 100.
Peñalara, 100.
Peñaranda, 100.
Peñasco, 98, 100.
Pêne, 102.
Peñol, 98, 100.
Peñoleria, 98, 100.
Peñon, 98.
Peu, 97.
Piano, 178.
Pic, 93, 94.
Picacho, 100.
Pico, 93, 94, 100.
Pié, 97.
Pike, 94, 95.
Pine Barren, 200.
Pique, 94, 102.
Piquette, 94, 102.
Piz, 96.
Pizzo, 96.
Plain, 178.
Plaine, 178.
Plains, 176, 179, 189.
Plateau, 180, 181.
Plateau Region, 180.
Pocosin, 211.
Poey, 102.

Port, 136.
Portezuelo, 141.
Portillo, 136.
Pouy, 97, 102.
Poy, 97.
Prairie, 188, 189, 190.
Prairie States, 189.
Precipice, 121.
Puch, 97.
Pueche, 97.
Puffing-hole, 172.
Puig, 97.
Puna, 195, 196.
Purgatory, 160–162.
Puy, 97.

QUAIRAT, 102.
Quebrada, 152.
Queyras, 102.
Queyre, 102.
Quiebra, 152.

RANGE, 87, 92.
Ravine, 149.
Reventazon, 101.
Reventon, 101.
Rocking-stone, 127.
Roque, 102.

SADDLE-BACK, 117.
Saracen's-stone, 128.
Sarrat, 102.
Sarsen, 128.
Savane, 185, 186.
Savanna, 183–188.
Scar, 123.
Scaur, 123.

Scaw, 123.
Scree, 124.
Scrogg, 227, 229.
Scrub, 227-229.
Sea, 80, 81.
Sea-shore, 80.
Serra, 88.
Serranía, 99.
Serrano, 99.
Serrata, 99.
Serrated, 134.
Serre, 102.
Shrub, 227.
Sierra, 88, 99.
Slack, 139.
Slashes, 203.
Soum, 102.
Spit, 96.
Spitze, 96.
Spitzli, 96.
Spur, 113.
Steppe, 192, 193, 219.
Stickle, 95.
Strait, 81.
Sty, 137.
Sugar-loaf, 117.
Swale, 212.
Swamp, 210, 211.
Swirl, 138.
Swirrel, 138.

Swyre, 138.
System, 92.

TABLE, 181.
Table-land, 180.
Table-mountain, 181.
Talus, 125.
Terrace, 182, 183.
Teton, 119.
Todi, 108.
Tooth, 116.
Tor, 126, 127.
Tower, 126.
Tundra, 199.
Tuque, 102.
Tuquet, 102.

VALE, 131.
Valle, 133.
Valley, 129-131.
Vigne, 102.

WASH, 125.
Water-gap, 135.
Water-shed, 141.
Weald, 213.
Wealden, 213.
Wind-gap, 135.
Wold, 216-218.

YOKE, 140.

University Press: John Wilson & Son, Cambridge.

www.ingramcontent.com/pod-product-compliance
Lightning Source LLC
Chambersburg PA
CBHW031752230426
43669CB00007B/583